Advances in Physical Organic Chemistry

ADVISORY BOARD

Advances in Physical Organic Chemistry

Volume 38

Edited by

J. P. RICHARD

Department of Chemistry
University of Buffalo, SUNY
Buffalo NY 14260-3000, USA

ACADEMIC PRESS

An imprint of Elsevier Science

Amsterdam Boston Heidelberg London New York Oxford Paris
San Diego San Francisco Singapore Sydney Tokyo

ELSEVIER SCIENCE Ltd
The Boulevard, Langford Lane
Kidlington, Oxford OX5 1GB, UK

First edition 2003

ISBN: 0-12-033538-7
ISSN: 0065-3160

Contents

Editor's preface

The speed of computers has increased exponentially during the past 50 years and there is no sense that an upper limit has been reached. This has resulted in a continuous assessment of the quality of the agreement between chemical experiments and calculations, and signs that the perpetual confidence of computational chemists in the significance of their calculations will eventually be fully justified, if this is not already the case. The interplay between computational and experimental chemists can be painful. It is sometimes diffcult for experimentalists to avoid the uncongenial and uncharitable view of computational chemists as dilettantes, with little interest in coming to grips with the tangled web of experimental work as needed to evaluate the agreement between theory and calculation and, consequently, no sense of the reactivity of real molecules and the mechanisms by which they react. Computational chemists may fee certain reservations regarding the abilities of experimentalists who become embroiled in interminable and unfathomable controversies about the interpretation of their data. It is understandable that they might view a world where experiments are rendered obsolete by computational infallibility as desirable. A degree of sympathy and mutual respect can be achieved through collaborations between experimental and computational chemists directed towards solving problems of common interest.

The question of the scope of Physical Organic Chemistry is often raised by those who recognize that this field is regarded by some as unfashionable, and who are concerned by the limited attention paid to problems that first spurred its development – Hammett relationships; reactive intermediates; proton-transfer at carbon; polar reaction mechanisms; and so forth. Those who identify with Physical Organic Chemistry have little choice but to work to expand its scope, while preserving a sense of coherence with earlier work. Computational chemistry is fully developed subdiscipline of chemistry; and, computational chemists who publish on problems of long-standing interest to physical organic chemists may shape reports of their work to emphasize either the computational methods, or the reactions being investigated. This monograph provides an audience for those who wish to report advances in physical organic chemistry that have resulted from well-designed computational studies.

Volume 38 of *Advances in Physical Organic Chemistry* is a testament to advances that can result through the thoughtful application of computational methods to the analysis of mechanistic problems not fully solved by experiment. It has been dedicated to Kendall Houk on the occasion of his 60th birthday by the chapter authors, former coworkers of Ken's who have written about problems of mutual interest. Ken's contributions to chemistry and his personality are recounted in opening remarks by Wes Bordon. In a broader sense, this volume recognizes the scope of Ken's contributions; and, his active mind and gracious personality which are central to an ability to convey a knowledge of Chemistry and an enthusiasm for its study to colleagues of all ages.

John P. Richard

Kendall N. Houk at Age 60

It is hard to believe that Ken Houk turned 60 on February 27, 2003. Ken continues eagerly to tackle new challenges, both professional and personal. As an example in the latter arena, last year Ken learned to ride a unicycle – a 59th birthday present from his wife Robin Garrell.

In addition, despite his magnificent contributions to chemistry and the many awards that he has won for them, Ken still has not learned to take himself seriously. This summer he and Robin convulsed an audience of quantum chemists by dressing and acting like movie stars on Oscar night when they presented the award for best poster at an international conference. People who meet Ken are amazed to discover that a chemist as famous as he can be so easy going and so funny. Nevertheless, Ken really is one of the people who helped to transform physical organic chemistry from the study of reaction mechanisms in solution to the much broader field that it is today.

Ken has been a leader in the development of rules to understand chemical reactivity and selectivity and in the use of computers to model complex organic and biological reactions. Ken's theoretical work has stimulated numerous experimental tests of predictions made by him, and some of these tests have been performed by his own research group. Ken has not only shown organic chemists how to use calculations to understand chemistry, but his papers and his lectures have also inspired experimentalists to use calculations in their own research.

Ken has published prolifically. He has authored or co-authored nearly 600 articles in refereed journals, an average of 10 papers/year since his birth in Nashville in 1943. The majority of his papers have appeared in *JACS*, but a smattering have been published in *Angew. Chem.*and in *Science*. Ken was the 35th most cited chemist in the world during the last two decades.

Ken has mentored nearly 150 graduate students, half that number of postdocs, and many times that number of undergraduates in his teaching career, first at LSU, then at Pittsburgh, and now at UCLA. Dozens of faculty members from other universities have spent sabbaticals in Ken's group, in order to work with and learn from Ken. Many of his students and postdocs are now themselves successful and distinguished scientists, as exemplified by the contributors to this volume.

In Ken are combined the physical insight of an organic chemist with the sophistication in computational methodology of a physical chemist. However, like Nobel Laureate Roald Hoffmann, less important than the quantitative results of Ken's calculations are the qualitative insights that have emerged from analyzing these results.

Ken's insights have shaped thinking in organic chemistry in many areas. The list of his contributions includes: theoretical models of reactivity and regio- and stereoselectivity in cycloadditions, the concerted nature of 1,3-dipolar and Diels-Alder reactions, the concept and theory of "periselectivity", the impossibility of "neutral homoaromaticity", the origin of negative activation energies in and entropy control of carbene addition reactions; the phenomenon and theoretical explanation of "torquoselectivity"; the origins of stereoselectivity in and practical methods for computational modeling of the transition structures of a wide variety of synthetically important reactions, gating in host-guest complexes, and mechanisms of transition state stabilization by catalytic antibodies. Many of the contemporary concepts that permeate organic chemists' notions of how organic reactions occur and why they give particular products originated in discoveries made in the Houk labs.

Like Roald Hoffmann and Ken's own Ph.D. adviser, R. B. Woodward, Ken seems to enjoy making up erudite-sounding names for new phenomena that he discovers. In addition to "periselectivity" and "torquoselectivity", Ken has added "theozyme" to the chemical lexicon.

In the beginning, Ken created a frontier molecular orbital (FMO) theory of regioselectivity in cycloadditions. In particular, his classic series of papers showed how FMO theory could be used to understand and predict the regioselectivity of 1,3-dipolar cycloadditions. Ken's generalizations about the shapes and energies of frontier molecular orbitals of alkenes, dienes, and 1,3-dipoles, are in common use today; and they appear in many texts and research articles.

In a very different area of organic chemistry Ken produced a series of landmark theoretical papers on carbene reactions. He developed a general theory, showing how orbital interactions influence reactivity and selectivity in carbene additions to alkenes. Ken also showed how entropy control of reactivity and negative activation barriers in carbene addition reactions could both be explained by a new, unified model.

With great insight, Ken pointed out that even if such reactions have vanishingly small enthalpic barriers, they still do involve very negative changes in entropy. The $-T\Delta S^{\ddagger}$ term in the free energy of activation produces a free energy barrier with an entropic origin. The position and height of this barrier both depend on how rapidly the enthalpy and entropy each decrease along the reaction coordinate and also on the temperature. Ken's theory has had a pervasive impact on the interpretation of fast organic reactions.

The name "Houk" has become synonymous with calculations on the transition states of pericyclic reactions. For two decades, as increasingly sophisticated types of electronic structure calculations became feasible for such reactions, Ken's group used these methods to investigate the geometries and energies of the transition structures. Ken's calculations showed that, in the absence of unsymmetrical substitution, bond making and bond breaking occur synchronously in pericyclic reactions.

In his computational investigations of electrocyclic reactions of substituted cyclobutenes, Ken discovered a powerful and unanticipated substituent effect on which of the two possible modes of conrotatory cyclobutene ring opening is preferred. He called this preference for outward rotation of electron donating substituents on the scissile ring bond "torquoselectivity." On this basis many unexplained phenomena were understood for the first time. The prediction that a formyl group would preferentially rotate inward, to give the less thermodynamically stable product, was verified experimentally by Ken's group at UCLA. The concept of torquoselectivity has blossomed into a general principle of stereoselection, and experimental manifestations of torquoselectivity continue to be discovered.

In a study of reactivity and stereoselectivity in norbornenes and related alkenes, the observation of pyramidalized alkene carbons led Ken to the discovery of a general pattern — alkenes with no plane of molecular symmetry pyramidalize so as to give a staggered arrangement about the allylic bonds. Subsequent studies showed that there is a similar preference for staggering of bonds in transition states.

Ken pioneered the modeling of transition states with force field methods. Before modern tools existed for locating transition structures in all but the simplest reactions, his group used *ab initio* calculations to find the geometries of transition states and to determine force constants for distortions away from these preferred geometries. These force constants could then be used in standard molecular mechanics calculations, in order to predict how steric effects would affect the geometries and energies of the transition structures when substituent were present.

Another series of publications from Ken's group compared kinetic isotope effects, computed for different possible transition structures for a variety of reactions, with the experimental values, either obtained from the literature or measured by Singleton's group at Texas A&M. These comparisons established the most important features of the transition states for several classic organic reactions — Diels-Alder cycloadditions, Cope and Claisen rearrangements, peracid epoxidations, carbene and triazolinedione cycloadditions and, most recently, osmium tetroxide bis-hydroxylations. Due to Ken's research. the three-dimensional structures of many transition states have become nearly as well-understood as the structures of stable molecules.

Ken has continued to explore and influence new areas of chemistry. For example, he has recently made an important discovery in molecular recognition. His finding that a conformational process ("gating") is the rate-determining step in complex formation and dissociation in Cram's hemicarceplexes has produced a new element in host design. Ken's investigations of the stabilities and mechanisms of formation of Stoddart's catenanes and rotaxanes have already led to discovery of gating phenomena in and electrostatic stabilization of these complexes.

Ken's calculations on catalytic antibodies provide a recent example of the fine way that he utilizes theory to reveal the origins of complex phenomena. His computations have led to the first examples of a quantitative understanding of the role of binding groups on catalysis by antibodies.

Ken's research has been recognized by many major awards. Among these some of the most significant are an Alexander von Humboldt U.S. Senior Scientist Award from Germany, the Schrödinger Medal of the World Association of Theoretically Oriented Chemists, the UCLA Faculty Research Lectureship, a Cope Scholar Award and the James Flack Norris Award of the American Chemical Society, the Tolman Award of the Southern California Section of the American Chemical Society, and an Honorary Degree ("Dr. honoris causa") from the University of Essen, Germany in 1999. In 2000, he was named a Lady Davis Professor at the Technion in Israel and received a Fellowship from the Japanese Society for the Promotion of Science. Last year Ken was elected to the American Academy of Arts and Sciences, and he has won the 2003 American Chemical Society Award for Computers in Chemical and Pharmaceutical Research.

Ken has gotten into his share of controversies. Among the most prominent of his sometime scientific adversaries have been Michael J. S. Dewar, Ray Firestone, George Olah, Fred Menger, Tom Bruice, and Arieh Warshel. However, Ken's sense of humor and refusal to take anything too seriously, including himself, has allowed him to remain good friends with (almost) all of these chemists at the same time they were having intense scientific disagreements.

Ken's long-term scientific friends outnumber his sometime scientific foes by at least two orders of magnitude. He has collaborated with an amazingly large number of the world's most outstanding chemists; and in my capacity as an Associate Editor of *JACS*, I have found that at least half of the organic theoreticians whose manuscripts I handle suggest Ken as a Referee. I am sure that they respect his critical judgement, but I suspect that they also believe that Ken is too nice a person to suggest that their manuscripts be rejected. Of course, I cannot possibly comment on whether or not they are right, but I can state that Ken unfailingly and promptly writes insightful reports on the comparatively small fraction of those manuscripts that I actually do send him.

However, Ken's service to the chemical community extends far beyond his willingness to referee promptly and thoroughly manuscripts that I send him. Ken has served as Chair of the Gordon Conferences on Hydrocarbon Chemistry and Computational Chemistry, two Reaction Mechanisms Conferences, and a recent Symposium honoring the life and chemistry of Donald Cram. He has also been Chair of the Chemistry and Biochemistry Department at UCLA, and for two years he was the Director of the Chemistry Division at the National Science Foundation.

I have known Ken for forty years, since we were both undergraduates at Harvard. He played trumpet in a jazz band, and I heard him perform on several occasions. I, as a Miles Davis wannabe (but one with no musical talent), noted with envy that, when Ken played, he adopted the same, highly characteristic posture as Miles. However, this was probably the last time in his life that Ken imitated anybody.

As Harvard graduate students, I with E. J. Corey and Ken with R. B. Woodward, we nodded politely at each other when we passed in the hall; but it was not until many years later, when we met at a conference, that I remember actually talking to Ken. In addition to both being theoretically inclined organic chemists, whose groups also did experiments, we discovered that we had other interests in common, interests which we still sometimes discuss but no longer pursue.

Through the years Ken and I have collaborated on several projects, all of them concerned with the Cope rearrangement. Some idea of the non-scientific side of Ken can be gleaned from his contributions to the late-night email messages we exchanged a few years ago in which the goal was to think of different words or phrases that incorporated "Cope" but had nothing to do with this pericyclic reaction. A few examples of Ken's creativity include "Cope ascetic", "Cope a cabana", and "Cope Ernie cuss".

However, I think Ken was at his creative best fifteen years ago when we coauthored an invited review on "Synchronicity in Multibond Reactions" for *Annual Reviews of Physical Chemistry*. This review was written to refute Michael Dewar's assertion in a *JACS* paper that "synchronous multibond reactions are normally prohibited". The review provided a rare occasion when Ken and I could each write on this subject without having to respond to a three-page, single-spaced, report from an "anonymous" Referee, which usually wound up by claiming that, if we weren't ignorant, then we must be scientifically dishonest in asserting that multibond reactions actually could be synchronous.

Given the freedom to include whatever we wished in this review, Ken suggested that we conclude with some comments on synchronicity from the non-scientific literature. Thus it was that our review ended with an excerpt from the song "Synchronicity" by Sting — "Effect without cause, Subatomic laws, Scientific pause, Synchronicity."

It has been my good fortune to know Ken for forty years as a friend, collaborator, and one of the most important and influential physical-organic chemists of the twentieth century. I have no doubt that, if Ken's unicycle does not put an untimely end to his brilliant career, his seminal contributions to chemistry will continue well into this century.

Wes Borden

Contributors to Volume 38

Jiali Gao Department of Chemistry and Supercomputing Institute, University of Minnesota, Minneapolis, Minnesota 55455, USA

Mireia Garcia-Viloca Department of Chemistry and Supercomputing Institute, University of Minnesota, Minneapolis, Minnesota 55455, USA

Jeehiun Katherine Lee Department of Chemistry, Rutgers, The State University of New Jersey, 610 Taylor Road, Piscataway, New Jersey, USA

Yirong Mo Department of Chemistry and Supercomputing Institute, University of Minnesota, Minneapolis, Minnesota 55455, USA

Jonas Oxgaard Department of Chemistry and Biochemistry, University of Notre Dame, Notre Dame, Indiana, USA

Michael N. Paddon-Row School of Chemical Sciences, University of New South Wales, Sydney, New South Wales, Australia

Tina D. Poulsen Department of Chemistry and Supercomputing Institute, University of Minnesota, Minneapolis, Minnesota 55455, USA

Nicolas J. Saettel Department of Chemistry and Biochemistry, University of Notre Dame, Notre Dame, Indiana, USA

Thomas Strassner Technische Universität München, Anorganisch-chemisches Institut, Lichtenbergstraße 4, D-85747 Garching bei München, Germany

Dean J. Tantillo Department of Chemistry and Chemical Biology, Cornell University, Ithaca, New York, USA

Olaf Wiest Department of Chemistry and Biochemistry, University of Notre Dame, Notre Dame, Indiana, USA

H. Zipse Department Chemie, LMU München, Butenandstr. 13, D-81377 München, Germany

Orbital interactions and long-range electron transfer

Michael N. Paddon-Row

School of Chemical Sciences, University of New South Wales, New South Wales, Sydney, NSW, 2052, Australia

Preamble

Professor Houk and I are coevals and we embarked on our research careers at about the same time. In the beginning of the 1970s, both he and I were independently working on mechanistic aspects of pericyclic reactions, using a combination of experiment, simple perturbational MO theory and semi-empirical MO calculations. My published work in this area was of variable quality whereas Ken's was uniformly

1

outstanding. I was due for sabbatical study in 1980. Although, by that time, I had begun my investigations into electron transfer, a story which is told below, I considered it daft not to spend 1980 with Ken, who was clearly on track to becoming one of the great American physical organic chemists. So I went to LSU in January, 1980, to become a member of "Houk's hordes", as they were then affectionately called. That year was the most rewarding, most exciting, and happiest year of my professional career. We did great work together, with Nelson Rondan, solving all sorts of challenging problems concerning π-facial stereoselectivity, using John Pople's spanking new GAUSSIAN 80 program that actually located stationary points automatically, thereby banishing for all time that dreadful axial iterative method for optimising molecular geometries! Over the intervening years, Ken and I have kept up our friendship and we even occasionally collaborate on projects of mutual interest. Although my interests have diverged somewhat from Ken's, I always read his papers for, like Roald Hoffmann's papers, Ken's are not only of the highest quality but they are also elegantly written. I respect and admire Professor Houk, not only for his chemical brilliance, but also for his humanity – his generosity, his great sense of humour, his liberal views and his tolerance of other people's points of view. So, I am absolutely delighted to have been invited to contribute an article to this volume in honour of Professor Houk's 60th birthday.

1   Introduction

This article is a semi-personal account of how we, and others, solved one of the outstanding problems in the electron transfer (ET) field, namely, the distance dependence of long-range, non-adiabatic, ET dynamics, and how this distance dependence varies with the nature and configuration of the medium between the redox couple (chromophores). There are two main reasons why the issue of the distance dependence of ET dynamics was (and continues to be) so significant. Firstly, ET is the most fundamental of all chemical reactions and is pervasive throughout chemistry and biology; it behoves us, therefore, to understand fully, the mechanistic characteristics of such a fundamental process, and the distance dependence of ET dynamics is a pivotal characteristic. Secondly, it has been known for some time that ET in proteins and DNA double helices may take place over very large distances, often exceeding 50 Å,[1] and so a detailed mechanistic knowledge of biological ET necessarily entails an understanding of its distance dependence.

There are three principal modes of ET, namely, thermal, optical and photoinduced ET, and these are shown schematically in Fig. 1. Optical ET differs from photoinduced ET in that ET in the former process results from direct electronic excitation into a charge transfer (CT) or intervalence band, whereas photoinduced ET takes place from an initially prepared locally excited state of either the donor or acceptor groups. Photoinduced ET is an extremely important process and it is widely studied because it provides a mechanism for converting photonic energy into useful electrical potential which may then be exploited in a number of ways. The most famous biological photoinduced ET reaction is, of course, that which drives

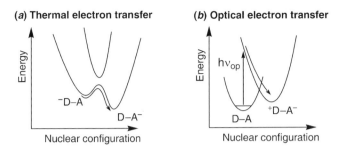

(a) Thermal electron transfer   (b) Optical electron transfer

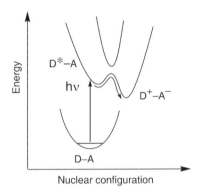

(c) Photoinduced electron transfer

Fig. 1   Three different types of electron transfer processes.

photosynthesis. Although these three types of ET processes appear to be quite distinct, they do share common fundamental mechanistic features, particularly regarding their distance dependence behaviour. Three important classes of ET are charge separation, charge recombination and charge-shift (Scheme 1) which may be promoted thermally, photochemically or optically. In the former two process, a dipole is created and destroyed, respectively, whereas in the charge-shift process, no dipole is created or destroyed.

This review is not intended to be exhaustive because two massive and outstanding tomes covering all aspects of ET have recently been published and the reader is directed to these works for an overview of the ET field.[2,3]

2   A simple theoretical model of ET

CLASSICAL THEORY

We begin by reviewing the elements of the classical theory of ET, developed independently by Marcus and Hush.[4,5] While more sophisticated theoretical treatments of ET exist,[6,7] Marcus–Hush theory will largely serve the purposes of

Charge separation

$$C_1 \ + \ C_2 \ \longrightarrow \ ^{\pm \bullet}C_1 \ + \ C_2^{\bullet \mp} \qquad \text{thermal}$$

$$^*C_1 \ + \ C_2 \ \longrightarrow \ ^{\pm \bullet}C_1 \ + \ C_2^{\bullet \mp} \qquad \text{photoinduced}$$

Charge recombination

$$^{\pm \bullet}C_1 \ + \ C_2^{\bullet \mp} \ \longrightarrow \ C_1 \ + \ C_2 \qquad \text{thermal}$$

$$^{\pm \bullet}C_1^* \ + \ C_2^{\bullet \mp} \ \longrightarrow \ C_1 \ + \ C_2 \qquad \text{photoinduced}$$

Charge-shift

$$^{\pm \bullet}C_1 \ + \ C_2 \ \longrightarrow \ C_1 \ + \ C_2^{\bullet \pm} \qquad \text{thermal}$$

$$^{\pm \bullet}C_1^* \ + \ C_2 \ \longrightarrow \ C_1 \ + \ C_2^{\bullet \pm} \qquad \text{photoinduced}$$

Scheme 1

this review. Consider an ET reaction in which a redox couple, **D–A**, gives the charge-separated (CS) state product $^+\text{D–A}^-$. The Marcus–Hush theory deconvolutes the potential energy surface for this reaction into two interacting parabolic diabatic potential energy curves, one representing the diabatic configuration of the reactant, **D–A**, and the other one representing the diabatic configuration of the product $^+\text{D–A}^-$, as illustrated in Fig. 2. The approximation of treating the diabatic energy surfaces as parabolas is a reasonable one, especially if free energies are used

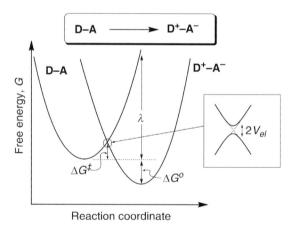

Fig. 2 Energy diagram for charge separation resolved into reactant-like and product-like diabatic surfaces. The reaction coordinate is a function of the generalised nuclear coordinates of the redox system and of the surrounding solvent molecules. The two diabatic curves do not, in general, intersect, but interact, to give an avoided crossing, whose energy gap is twice the electronic coupling, V_{el}, for the interaction.

instead of potential energies.[5] The reaction coordinate for ET includes changes in both the geometry of the **D**–**A** system and solvent orientation which accompanies the reaction. The two diabats formally cross one another at some point along the one-dimensional reaction coordinate. In this crossing region, the two configurations mix, symmetry permitting, and this results in an avoided crossing whose magnitude is equal to $2V_{el}$, where V_{el} is the electronic coupling matrix element and may be regarded as a rough measure of the strength of the interactions between the MOs of **D** and those of **A**.

Within the context of transition state theory, the ET rate, k_{et}, may be expressed by equation (1) in which k_B and T are the Boltzmann constant and absolute temperature, respectively, and $\Delta G^{\ddagger}$ is the free energy of activation, taken to be the height of the avoided crossing above the reactant well (Fig. 2).

$$k_{et} = \tau_{el} \nu_n \exp\left(\frac{-\Delta G^{\ddagger}}{k_B T} \right) \tag{1}$$

The remaining two terms in equation (1), ν_n and τ_{el} are, respectively, the nuclear frequency factor and the electronic transmission coefficient. The frequency factor gives the frequency with which reaction trajectories reach the avoided crossing region, and τ_{el} gives the probability that, once a trajectory has reached the avoided crossing region, it will pass into the product well, rather than be deflected back into the reactant well.

The simple geometric properties of parabolas enable the free energy of activation to be expressed as $\Delta G^{\ddagger} = (\Delta G^0 + \lambda)^2 / 4\lambda$, in which case, equation (1) may be rewritten as equation (2), generally called the classical Marcus equation:

$$k_{et} = \tau_{el} \nu_n \exp\left(-\frac{(\Delta G^0 + \lambda)^2}{4\lambda k_B T} \right) \tag{2}$$

In this equation, ΔG^0 is the free energy change associated with the ET process and λ is the so-called reorganisation energy. The reorganisation energy is defined as the energy required to distort the reactant and its associated solvent molecules, from their relaxed nuclear configurations, to the relaxed nuclear configurations of the product and its associated solvent molecules (Fig. 2). It is common practice to represent the total reorganisation energy, λ, as the sum of a solvent independent internal term, λ_{int}, (inner sphere), arising from structural differences between the relaxed nuclear geometries of the reactant and product, and a solvent reorganisation energy term, λ_s (outer sphere), arising from differences between the orientation and polarisation of the solvent molecules surrounding the reactant and product.[8] Thus, $\lambda = \lambda_{int} + \lambda_s$.

The great utility of the Marcus equation (2) lies in recognising that the free energy changes, ΔG^0, for most ET processes are exergonic, i.e., they are negative quantities, whereas the reorganisation energies are always positive quantities. Consequently, for the case where $-\Delta G^0 < \lambda$, termed the Marcus normal region, the ET rate increases with increasing exergonicity – because $(\Delta G^0 + \lambda)^2$ is decreasing – and finally reaches a maximum (optimal) value when $-\Delta G^0 = \lambda$, at which point the process becomes activationless (Fig. 3). As the reaction becomes more exergonic, the ET process enters the Marcus inverted region. In this region, the activation barrier reappears – because $(\Delta G^0 + \lambda)^2$ is always positive – and the ET rate accordingly decreases with increasing exergonicity. This interesting predicted dependence of ET rate on exergonicity, illustrated schematically in Fig. 3, was elegantly confirmed experimentally for thermal intra-molecular charge-shift reactions occurring in the radical anions of the rigid **D–steroid–A** bichromophoric molecules (dyads), which were generated by pulse radiolysis (Fig. 4).[9] The driving force $(-\Delta G^0)$ for the reaction was adjusted by changing the acceptor. It was found that, in qualitative agreement with classical Marcus theory, the ET rate increased with increasing exergonicity, peaked and then fell off. Note that the rigid hydrocarbon bridge – a steroid in this case – serves to maintain constant donor–acceptor separation in the **D–steroid–A** series.

ADIABATIC AND NON-ADIABATIC ET

From a mechanistic point of view, it is useful to divide ET into two classes which differ in the magnitude of the electronic coupling matrix element, V_{el}. Those ET processes for which $V_{el} > 200 \, \text{cm}^{-1}$ (2.4 kJ/mol) are called adiabatic because they take place solely on just one potential energy surface, namely that which connects the reactant and product states (i.e., see the dashed arrow in Fig. 5a). In this case, the large majority of reaction trajectories that reach the avoided crossing region will

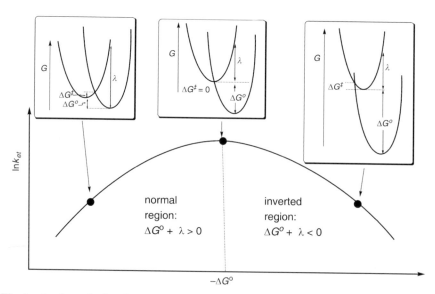

Fig. 3 A schematic showing how, within the context of classical Marcus theory, the ET rate varies with the ergonicity – or, equivalently, the driving force $(= -\Delta G^0)$ – of the reaction.

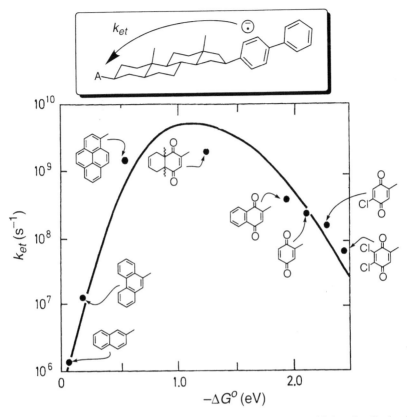

Fig. 4 The dependence of the rate of thermal ET taking place from a biphenyl radical anion donor to a series of acceptor groups **A**, as a function of driving force.[9]

successfully pass over into the product well (these trajectories are depicted as solid arrows in Fig. 5a). In other words, the electronic transmission coefficient, τ_{el}, in equations (1) and (2) is unity.

In contrast, those ET processes for which V_{el} is much less than $200 \, \text{cm}^{-1}$ are termed non-adiabatic because a significant majority of reaction trajectories which enter the avoided crossing region undergo non-adiabatic quantum transitions (i.e., surface hops) to the upper surface (curved arrows in Fig. 5b), and only a small fraction of trajectories remain on the lower surface and lead to ET (straight arrow in Fig. 5b). Those trajectories which undergo a non-adiabatic transition to the upper surface are reflected by the right-hand wall of the upper surface and re-enter the avoided crossing region where they are likely to undergo another non-adiabatic transition, but this time, to the lower surface. However, the conservation of momentum dictates that these trajectories will proceed in the direction of the reactant well, rather than towards the product well, where they will be reflected by the left-hand wall of the reactant well and re-enter the avoided crossing

(*a*) Adiabatic electron transfer   **(*b*) Nonadiabatic electron transfer**

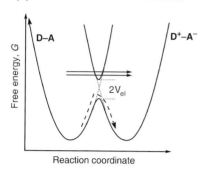

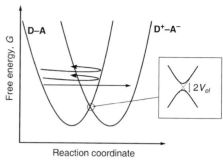

$$V_{el} > 200 \text{ cm}^{-1} \implies \tau_{el} \approx 1 \qquad\qquad V_{el} < 200 \text{ cm}^{-1} \implies \tau_{el} < 1$$

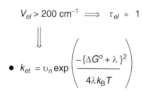

$$\bullet\; k_{et} = \upsilon_n \exp\left(\frac{-\{\Delta G^{\circ} + \lambda\}^2}{4\lambda k_B T}\right) \qquad \bullet\; k_{et} = \frac{4\pi^2 |V_{el}|^2}{h}\left\{\frac{1}{4\pi\lambda k_B T}\right\}^{1/2} \exp\left(\frac{-\{\Delta G^{\circ} + \lambda\}^2}{4\lambda k_B T}\right)$$

Fig. 5 Adiabatic and non-adiabatic ET processes. In the adiabatic process (Fig. 5a), $V_{el} > 200 \text{ cm}^{-1}$ and the large majority of reaction trajectories (depicted as solid arrows) which reach the avoided crossing region remain on the lower energy surface and lead to ET and to the formation of product (i.e., the electronic transmission coefficient is unity). In contrast, non-adiabatic ET is associated with V_{el} values $< 200 \text{ cm}^{-1}$, in which case the majority of reaction trajectories which reach the avoided crossing region undergo non-adiabatic transitions (surface hops) to the upper surface. These trajectories rebound off the right-hand wall of the upper surface, enter the avoided crossing region where they are likely to undergo a non-adiabatic quantum transition to the lower surface. However, the conservation of momentum dictates that these trajectories will re-enter the reactant well, rather than the product well. Non-adiabatic ET is therefore associated with an electronic transmission coefficient which is less than unity.

region, where they will attempt, yet again, to undergo (adiabatic) ET by remaining on the lower surface. The upshot of this is that non-adiabatic reaction trajectories take longer to reach the product well, than adiabatic trajectories, with the result that the overall rate of non-adiabatic ET is slowed down. In other words, τ_{el} for non-adiabatic ET is much less than unity, its magnitude varying with the square of V_{el}. Therefore, the non-adiabatic ET rate also depends on the square of V_{el} and, within the context of the Fermi Golden Rule, which applies to the weak electronic coupling (i.e., non-adiabatic) limit, it takes the form of equation (3):

$$k_{et} = \frac{4\pi^2}{h} |V_{el}|^2 \text{FCWD} \tag{3}$$

where FCWD is the Franck–Condon weighted density of states which contains information concerning the overlap integrals between the reactant and product vibrational wavefunctions. In the Marcus semi-classical treatment, equation (3) is approximated by equation (4).[4,5]

$$k_{et} = \frac{4\pi^2 |V_{el}|^2}{h} \frac{1}{\sqrt{4\pi\lambda k_B T}} \exp\left(\frac{-[\Delta G^0 + \lambda]^2}{4\lambda k_B T}\right) \qquad (4)$$

Thus, the semi-classical Marcus theory of non-adiabatic ET expresses the ET rate constant in terms of three important quantities, namely V_{el}, λ, and ΔG^0. It therefore follows that an understanding of ET reactions entails an understanding of how these three variables are dependent on factors such as the electronic properties of the donor and acceptor chromophores, the nature of the intervening medium and the inter-chromophore separation and orientation.

In general, adiabatic ET holds sway when the donor and acceptor chromophores are in van der Waals contact, in which case V_{el} should be much larger than 200 cm^{-1}. Consequently, non-adiabatic ET rates are independent of the magnitude of V_{el}. Non-adiabatic ET, on the other hand, should dominate for separations greatly exceeding the sum of the van der Waals radii of the donor and acceptor, in which case the term long-range ET is reserved. In this instance, the magnitude of V_{el} is very small and, therefore, long-range ET rates are strongly dependent on the magnitude of V_{el}. It is reasonable to state that a detailed understanding of long-range ET necessarily entails an understanding of those factors which determine the magnitude and distance dependence of V_{el}. Long-range ET is of considerable interest because of its prevalence in many important biological systems, such as the photosynthetic reaction centre.[10-13] For the remainder of this article, attention will mainly focus on aspects of long-range (i.e., non-adiabatic) ET.

3 The distance dependence problem of non-adiabatic ET

THROUGH-SPACE DISTANCE DEPENDENCE

The distance dependence of ET dynamics is influenced by the distance dependence characteristics of the three parameters which appear in the Marcus equation (4), namely, V_{el}, λ, and ΔG^0. Although the driving force $(-\Delta G^0)$ and the reorganisation energy, λ – especially the solvent reorganisation energy, λ_s – usually display a distance dependence, it is often weak and may even be turned off under certain experimental conditions. For example, λ_s for non-polar solvents is very small, often less than 0.05 eV (5 kJ/mol), and so the distance dependence of λ_s is likewise negligible. The driving force for charge-shift ET processes (Scheme 1) displays little, if any, distance dependence because this type of ET reaction does not involve charge separation, and hence there is no accompanying change in electrostatic interactions. In contrast to the behaviour of ΔG^0 and λ_s, the electronic coupling element, V_{el}, is strongly distance dependent, and, in general, it is this variation of V_{el} which mainly governs the distance dependence of non-adiabatic ET rates – see equations (3) and (4). The strength of V_{el} is expected to decay exponentially with increasing donor–acceptor separation. This follows from the definition of V_{el}, given by equation (5). The matrix element, involving the interaction between the reactant diabatic wavefunction, Ψ^0_{DA}, and the product diabatic wavefunction, $\Psi^0_{D^+A^-}$, may be approximated by the one-electron integral involving the two active orbitals, i.e., the donor MO, φ_D, which initially contains the migrating electron, and the acceptor MO, φ_A, which receives the migrating electron. This integral depends on the overlap integral between φ_D and φ_A and these types of integrals are well known to decrease exponentially with increasing inter-orbital separation.

$$V_{el} = \langle \psi^0_{DA} | \hat{H} | \psi^0_{D^+A^-} \rangle \approx \langle \varphi_D | \hat{h}_{(1)} | \varphi_A \rangle \tag{5}$$

Thus, both V_{el} and the associated ET rate constant, k_{et}, should decay exponentially with increasing inter-chromophore separation, r, according to equations (6) and (7):

$$V_{el} \propto \exp(-0.5\beta_{el}r) \tag{6}$$

$$k_{et} \propto \exp(-\beta r) \tag{7}$$

where β_{el} and β are damping factors. It is often assumed that β_{el} and β have identical magnitudes, but this is not strictly correct because, as discussed above, β, being a phenomenological quantity, incorporates distance dependence contributions, not only from V_{el}, but also from Franck–Condon factors, such as the driving force and the solvent reorganisation energy (see equations (3) and (4)). Thus, β is expected to be slightly larger than β_{el}. In this article, we shall continue to use the symbols β_{el} and β to denote the damping factors associated with the distance dependence of V_{el} and the ET rate, respectively.

Let us first consider the case of ET taking place in a vacuum. In the absence of any intervening medium between the redox couple, the electronic coupling depends on the direct, through-space (TS) overlap between the active orbitals φ_D and φ_A of the donor and acceptor groups, respectively. The distance dependence of TS coupling may be estimated by calculating the coupling between the π MOs of two ethene molecules, lying in parallel planes, as a function of inter-orbital separation (Fig. 6). At infinite separation, the two π MOs are degenerate. At closer distances, the degeneracy is lifted by TS orbital overlap and gives rise to a splitting energy, $\Delta E(\pi)$, which is taken to be a positive quantity. It is readily shown that, to a good approximation, $\Delta E(\pi)$ is twice the magnitude of the electronic coupling matrix element, V_{el}, for hole transfer in the radical cation of the ethene dimer.[14] A simple *ab initio* MO calculation on the system shown in Fig. 6 predicts an exponential decay in the magnitude of the TS splitting energy $\Delta E(\pi)$, with $\beta_{el} = 2.8 - 3.0$ Å^{-1} (from HF/6-31 + G and HF/4-31G calculations).[14] This is a large value for the damping factor and it suggests that the ET rate occurring by a TS mechanism could be attenuated by a factor as large as 20 per Å. Consequently, long-range ET occurring TS is expected to be quite slow for donor–acceptor separations exceeding 6 Å.

Through–space (TS) coupling:

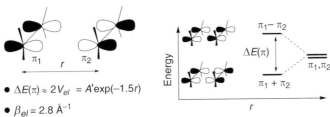

$$\Delta E(\pi) \approx 2V_{el} = A'\exp(-1.5r)$$

$$\beta_{el} = 2.8 \text{ Å}^{-1}$$

Fig. 6 A schematic of TS orbital interactions between the π MOs of two ethene molecules approaching each other in parallel planes. At infinite separation, the two MOs are degenerate. At closer distances, the degeneracy is lifted by an amount $\Delta E(\pi)$, the so-called splitting energy, which may be equated to twice the magnitude of the electronic coupling matrix element, V_{el}, for hole transfer in the radical cation of the ethene dimer. The β_{el} value was obtained from HF/6-31 + G calculations.[14]

THE INFLUENCE OF THE MEDIUM

In order to investigate the distance dependence of ET dynamics experimentally, it is advisable to employ donor–acceptor systems that are "rigid" in the sense that the distance and orientation between the donor and acceptor groups are well-defined and are subject to controlled and systematic variation. It is possible to investigate the distance dependence of ET dynamics inter-molecularly by dissolving known amounts of the donor and acceptor species in glasses. However, in these systems, the donor and acceptor distances and orientations are statistical distributions, although reasonable ET rate-distance profiles have been obtained from statistical analysis of the rate data.[15,16] A more tractable method is to use systems in which the donor and acceptor groups are connected to each other, either by covalent linkages or by non-covalent interactions, such as H-bonds or electrostatic interactions. Covalently linked donor–acceptor systems have provided the most detailed and unambiguous insight into the distance dependence of long-range ET. These systems will be generically denoted by **D–B–A** for a bichromophoric system (or dyad), in which the donor (**D**) and acceptor (**A**) groups are covalently attached to a bridge (**B**).

There is another reason for studying the distance dependence of both intra-molecular ET dynamics in covalently linked **D–B–A** systems, and inter-molecular ET dynamics in systems in which there is some kind of intervening medium between the donor and acceptor groups, and this involves the question of whether this distance dependence is influenced by the nature of the intervening bridge or medium. In other words, does the bridge serve purely as an inert spacer, whose sole purpose is to hold the donor and acceptor groups in a well-defined separation and orientation, or does it actively participate in the ET process? Clearly, the intervening medium could influence ET dynamics by its effect on both the driving force $(-\Delta G^0)$ and the solvent reorganisation energy terms in the Marcus equation (4) but this influence is qualitatively well-understood. More importantly, the intervening medium – hereafter referred to as the bridge, irrespective of whether it is or is not covalently linked to the redox couple – might significantly influence ET rates by modulating the strength of V_{el}, equation (5), through mixing of the bridge wavefunction with the reactant and product diabatic wavefunctions.

Just as exponential behaviour is predicted for the TS distance dependence of V_{el} it is reasonable to expect similar exponential behaviour if the vacuum is replaced by some kind of material, although the magnitude of the damping factor, β_{el}, might well be different from that arising from a TS coupling mechanism. Although this behaviour is often observed, the actual mechanism and the associated distance dependence characteristics of bridge-mediated ET depend, in part, on the magnitude of the energy gaps between the virtual states of the intervening bridge and the donor and acceptor states or, in simplistic MO terminology, they depend on the energy gaps between the donor HOMO and the bridge LUMO, and between the acceptor LUMO and the bridge HOMO. Other contributing factors to the distance dependence of bridge-mediated ET include the strengths of the matrix elements of interaction between the bridge wavefunction and the donor and acceptor wavefunctions, and the strength of the coupling between component elements of the bridge (intra-bridge coupling), and these will be discussed later on; but first, we shall investigate the influence of the chromophore–bridge energy gap on the ET mechanism.

ELECTRON TRANSFER AND ELECTRON TRANSPORT

Fig. 7 illustrates the influence of this energy gap using a variety of bridges. If the HOMO(donor)–LUMO(bridge) energy gap is very small, or non-existent (i.e., $\leq k_B T$), then electron migration from the donor group to the acceptor group takes place by an electrical conduction process, i.e., by an incoherent, inelastic scattering mechanism.[17–23] If the bridge is some kind of metallic nanoparticle (Fig. 7a, left-hand side), then the electron conduction through the bridge takes place by inelastic electron-lattice scattering (ohmic scattering). The energy gap between the conduction and valence bands of the metal nanoparticle is narrow, but not

---

Fig. 7   Three different modes of bridge behaviour towards charge migration processes. In the energy level diagrams (insets), the filled (black) strips and boxes denote filled levels and the hollow strips and boxes denote vacant (virtual) levels. (a) The donor HOMO and the acceptor LUMO lie within the bridge's valence and conduction bands, respectively. In metallic conduction, the soliton band is absent from the energy diagram, but is generally present in the case of conduction through conjugated organic polymers. The charge migration processes described in (a) are called charge transport processes. (b) Conduction takes place by charge hopping which occurs between adjacent, discrete sites of similar energies. This is another example of charge transport. (c) Charge migration takes place coherently by a superexchange mechanism. This type of charge migration is called electron (or hole) transfer.

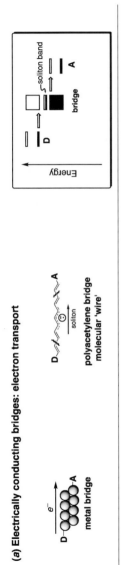

(a) Electrically conducting bridges: electron transport

metal bridge

polyacetylene bridge
molecular 'wire'

soliton band

bridge

(b) Charge hopping bridges: electron transport

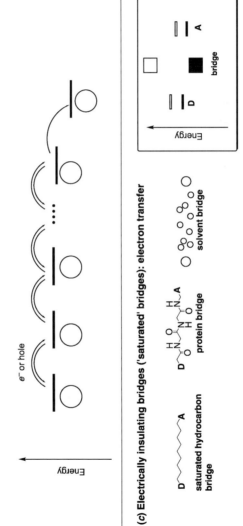

(c) Electrically insulating bridges ('saturated' bridges): electron transfer

saturated hydrocarbon
bridge

protein bridge

solvent bridge

bridge

necessarily zero, as it is in the macroscopic metal. The important point to note here is that the HOMO of the donor (either in its ground or electronically excited state) lies energetically within the metal conduction band, as shown in the right-hand inset of Fig. 7a, in which the filled (black) strips and boxes denote occupied levels, and the unfilled (white) strips and boxes denote vacant (virtual) levels (ignore the soliton band for the time being). The acceptor LUMO acts as a drain and must lie energetically within the metal conduction band.

If the bridge comprises an extended conjugated polymer, which may, or may not be doped, then electrical conduction occurs by bridge defects (i.e., solitons, polarons, bipolarons, etc.) which are propagated along the bridge by a soliton–phonon scattering mechanism.[17,24,25] If a one-dimensional conjugated polymer bridge (often called a molecular wire) is doped, then a soliton band is present, generally appearing in the mid-gap region of the polymer (Fig. 7a). Again, the donor HOMO must lie energetically close to either the polymer's conduction band or the mid-gap soliton band (for doped polymers). For these two conduction processes – metallic conduction and soliton conduction in conjugated polymers – the distance dependence of the electron (or hole) migration rate is ohmic, i.e., it varies inversely with bridge length; under certain conditions it is predicted that the ET or hole migration rate may even display no distance dependence at all.[19]

A third type of electrical conduction that may take place through certain bridges is called charge-hopping.[26–31] In this process, the bridge comprises a set of weakly coupled discrete units (B_i), possessing similar redox characteristics, each unit being able to capture the migrating charge for a short period of time, before passing it on to one of its neighbouring units (Fig. 7b). For example, the bridge might consist of a series of aromatic groups which may or may not be connected to each other. The migrating charge (electron or hole) randomly hops up and down the chain, until it is eventually irreversibly trapped by the acceptor which acts as a thermodynamic sink. If the bridge units are equally spaced, then the hopping dynamics will display a weak distance dependence, varying inversely with respect to a small power of the number, N, of hopping steps, namely, equation (8):

$$k_{et} = N^{-\eta} \qquad \eta \approx 1-2 \tag{8}$$

It is important to realise that, for all three types of electrical conduction processes discussed earlier, the migrating electron (or hole) is injected into the bridge, and that it moves through the bridge; therefore, it becomes localised within the bridge. This type of electron or hole migration, in which the particle moves through the bridge, is termed electron transport, in order to distinguish it from ET, in which the transferred electron is never localised within the bridge (see later).

In summary, the distance dependence of electron transport dynamics varies as r^{-p}, where r is the bridge length (or the number of bridge units in the case of a charge hopping process) and p is less than 2. Consequently, electron transport rates display a very weak dependence on bridge length.

If the energy gaps between the bridge and chromophore states are large (i.e., several eV), as is the case for bridges comprising saturated hydrocarbon moieties, proteins and solvent molecules (Fig. 7c), then electrical conduction cannot take place within these systems under normal conditions (i.e., low applied voltages). In this case, the bridge may, nevertheless, facilitate the electron migration process by another mechanism, known as superexchange.[32–35] In crude terms the super-exchange mechanism refers to the interaction between the orbitals of the bridge (π, π^*, σ, σ^*, etc.) and those of the donor and acceptor groups. In essence, the bridge provides a tunnelling pathway for the migrating electron. Consequently, the migrating electron is never resident within the bridge; thus, superexchange-mediated electron migration is a type of ET (as opposed to electron transport, see earlier). For those cases of superexchange-mediated ET where the bridge is saturated and only σ and σ^* orbitals are available for coupling with the chromophore MOs, the superexchange mechanism is often referred to as a through-bond (TB) coupling mechanism or orbital interactions through bonds.[14,36–39]

THE SUPEREXCHANGE MECHANISM

The superexchange mechanism is schematically depicted in Fig. 8 for the case of charge separation occurring in a **D**–**B**–**A** system in which the bridge, **B**, consists of *n* subunits, B_i, which might, for example, be two-centre bonds, such as C–C bonds, or larger entities, such as aromatic rings or DNA base pairs. The charge separation process is regarded as the coupling of the reactant **D**–**B**–**A** and product $^+$**D**–**B**–**A**$^-$ diabats with the two sets of virtual ionic bridge configurations $^+$**D**–**B**$_i^-$–**A** and **D**–**B**$_i^+$–**A**$^-$ ($i = 1, 2, ..., n$) which are generated from CT excitations from the donor to the bridge, in the former set, and from the bridge into the acceptor, in the latter set.

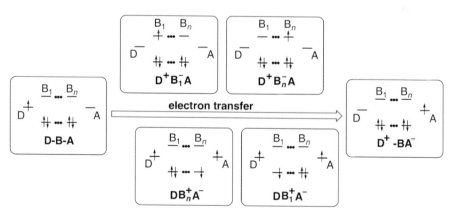

Fig. 8 In the superexchange mechanism, virtual states resulting from ionisation of the bridge mix into the reactant and product diabatic configurations to give the coupling, V_{el}. If virtual anionic bridge states, $^+$**D**–**B**$^-$–**A** are more important than cationic bridge states **D**–**B**$^+$–**A**$^-$ in the mixing scheme, the process is termed electron transfer and if the reverse holds, then it is termed hole transfer.

These virtual ionic bridge configurations have no physical existence; they are not intermediates in the charge separation process since their energies are much higher than those of the reactant and product states. The mixing of the reactant $\mathbf{D-B-A}$ and product $^+\mathbf{D-B-A}^-$ configurations with the virtual ionic bridge configurations leads to an increase in the strength of the electronic coupling, V_{el}, for charge separation over that resulting merely from a direct, TS interaction between $\mathbf{D}$ and $\mathbf{A}$. If the $^+\mathbf{D-B}_i^--\mathbf{A}$ configurations mix more strongly with the reactant and product configurations than do the $\mathbf{D-B}_i^+-\mathbf{A}^-$ configurations, then the charge separation is said to occur by an ET mechanism because the dominant interactions are between the HOMOs of the donor and the LUMOs of bridge. If, on the other hand, the $\mathbf{D-B}_i^+-\mathbf{A}^-$ configurations mix more strongly with the reactant and product configurations than the $^+\mathbf{D-B}_i^--\mathbf{A}$ configurations, then charge separation is said to occur by a hole transfer (HT) mechanism, because the dominant interactions are between the acceptor LUMOs and the bridge HOMOs.

A pictorial description of the superexchange mechanism may be obtained by inspecting the spatial distributions of the donor HOMO and the acceptor LUMO in a $\mathbf{D-B-A}$ system, such as that depicted in Fig. 9.[40,41] This dyad consists of a benzoquinone acceptor and an aniline donor, each fused to the ends of a rigid norbornylogous bridge, six bonds in length. The aniline HOMO and the benzoquinone acceptor LUMO are the active MOs which participate in optical ET in this molecule (see Fig. 1b). The geometry of the saturated hydrocarbon bridge is such as to allow hyperconjugative (through-bond) mixing of the donor and acceptor π and π^* MOs with the bridge σ and σ^* MOs. Indeed, as may be seen from Fig. 9, neither the donor HOMO nor the acceptor LUMO is localised within each chromophore; instead both MOs extend into the bridge, their amplitudes decaying exponentially with increasing penetration into the bridge. This orbital extension arises from the admixture into the donor HOMO and the acceptor LUMO small amounts of the bridge σ and σ^* MOs. The overlap integral between the donor HOMO and the acceptor LUMO is therefore enhanced by this TB-induced orbital extension mechanism and this leads to an enhanced magnitude of V_{el}. This coupling would be negligible in this dyad if all interactions of the chromophores with the bridge were switched off.

A simple perturbational orbital treatment of the superexchange mechanism was developed by McConnell,[34,39] and is illustrated in Fig. 10, for the case of through-bond-mediated ET occurring in a dyad comprising two identical chromophores covalently linked to a pentamethylene bridge. The bridge units are the individual C–C bridge bonds. In the simple McConnell scheme, the donor and acceptor chromophores each contribute a single π orbital to the interaction and each C–C bridge bond which is included in the treatment contributes a single σ or σ^* MO. The chromophores' orbitals, π_1 and π_2, are coupled to the allylic C–C σ MOs by matrix elements denoted by T (Fig. 10b), and the interactions between nearest neighbour bridge σ MOs are denoted by t. An important approximation made in the original McConnell treatment[34] is that all intra-bridge coupling interactions are restricted to nearest neighbour ones, i.e., t.

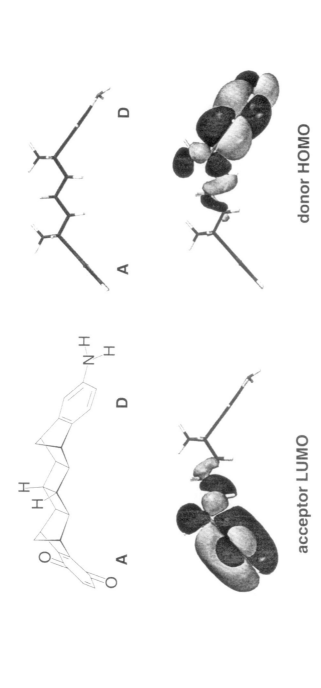

Fig. 9 The HF/3-21G HOMO and LUMO of the rigid benzoquinone-6-aniline donor acceptor system. The HOMO is associated with the aniline donor and the LUMO with the benzoquinone acceptor. These MOs are the active orbitals involved in optical ET in this molecule (see Fig. 1b). Note that the benzoquinone LUMO is not entirely localised within this group, but extends into the bridge, by a hyperconjugation mechanism, the LUMO amplitude decaying exponentially with increasing penetration into the bridge. This type of orbital extension is also observed for the aniline HOMO.

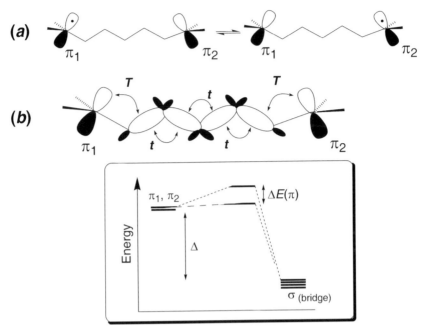

Fig. 10 Orbital description of the McConnell superexchange (through-bond) mechanism applied to a system comprising a donor and an acceptor covalently linked to a pentamethylene chain. The donor and acceptor chromophores each contribute a single π orbital to the interaction and each C–C bridge bond is assumed to contribute a single σ or σ^* MO, depicted as the former type in the figure. T is the interaction matrix element between a chromophore π MO and the allylic C–C σ MO, and t is the interaction matrix element between two *geminal* C–C σ MOs.

In this system, the predominant coupling involves the bridge σ orbitals, rather than the σ^* orbitals, because the former are closer in energy to the donor level than are the latter.[38] Mixing of the degenerate pair of chromophore π orbitals with the σ manifold of the bridge results in the lifting of the π orbital degeneracy by an amount $\Delta E(\pi)$ which, to second order, is given by equation (9):

$$\Delta E(\pi) = 2V_{el} = -2\left(\frac{T^2}{\Delta}\right)\left(\frac{t}{\Delta}\right)^{n-1} \tag{9}$$

where Δ is the energy gap between the chromophore π orbitals and the bridge σ orbitals prior to their interaction with each other, and n is the number of σ orbitals (= 4, for the dyad depicted in Fig. 10). The splitting energy, $\Delta E(\pi)$, is equal to twice the electronic coupling element, V_{el}, for the ET process. The perturbation expression shown in equation (9) is quite intuitive because $\Delta E(\pi)$ is resolved into the product of two types of couplings, one giving the coupling of the chromophores to the bridge (T^2/Δ), and a t/Δ factor for each interaction between adjacent bridge sites (Fig. 10b). The intra-bridge coupling term, $(t/\Delta)^{n-1}$, describes the propagation of the interaction along the bridge. An important requirement for this perturbation treatment to be valid is that the absolute magnitude of t/Δ is much less than unity, and it was assumed to be about 0.1 in the original treatment.[34]

It is immediately apparent from equation (9) that the McConnell model predicts an exponential decay of the magnitude of V_{el} with increasing number of bridge bonds. From equations (6) and (9) – using n, the number of bonds, in place of the distance r, in the former equation – the damping factor β_{el} may be expressed in terms of Δ and t according to equation (10):

$$\beta = 2 \ln \left| \frac{\Delta}{t} \right| \tag{10}$$

This treatment may be extended to cover the more general case of non-degenerate chromophores and non-identical bridge sites,[7,22,42–45] but the overall conclusion derived from the simple treatment given above still holds, namely that the electronic coupling decays exponentially with increasing bridge length.

In summary, both TS and superexchange (or TB) mechanisms predict an exponential decay of V_{el} with increasing donor–acceptor distance. The β_{el} value for the TS coupling mechanism is about 2.8–3.0 Å^{-1}, which suggests a sharp attenuation in the ET rate with increasing distance, amounting to a 20-fold reduction per Å. In his original treatment, McConnell suggested that the superexchange-mediated electronic coupling through a saturated hydrocarbon chain might have a β_{el} value as large as 2.3 bond^{-1} (i.e., $|t/\Delta| \approx 0.1$), although it was stated[34] that the numerical details should not be taken too seriously. A β_{el} value of 2.3 bond^{-1} for TB-mediated electronic coupling translates into a 10-fold attenuation in the corresponding ET rate per C–C bond. This is still a pretty steep distance dependence, even though it is smaller than that for TS-mediated coupling. For example, extending a hydrocarbon bridge by six bonds is predicted to result in a 1000-fold reduction in the magnitude of V_{el} and a million-fold reduction in the corresponding ET rate. This somewhat pessimistic prediction of the low efficacy of saturated hydrocarbon bridges to participate in superexchange (TB) interactions seemed reasonable at the time, in light of the known large energy gap existing between the π (and π^*) levels of typical donor and acceptor groups and the σ (and σ^*) MOs of saturated hydrocarbon bridges.

4　Experimental investigations of superexchange-mediated ET

SATURATED HYDROCARBON BRIDGES

A huge literature now exists describing experimental investigations of ET dynamics in D–B–A dyads, in which the donor and acceptor groups are covalently linked to a saturated hydrocarbon tether, or bridge, and only a small, but representative number of studies will be discussed here.[41,46] Why is there so much interest in saturated hydrocarbon bridges? The original motivation was to investigate the TS distance dependence of ET dynamics and it was thought that the best way to do this was to covalently attach the donor and acceptor groups to an inert bridge of variable length. The mindset during the 1960s and 1970s was that saturated hydrocarbon bridges are ideal inert spacers, on account of their predicted weak participation in superexchange-mediated electronic coupling. Indeed, hydrocarbon bridges were – and, sometimes, still are – referred to as "spacers" in ET investigations, which underscores the notion that these types of bridges serve merely to fix the chromophores in well-defined orientations and separations and that they were non-participating entities in the ET processes.

Early investigations of the distance dependence of ET dynamics focused on intra-molecular electron-spin transfer reactions in radical anions containing two identical aromatic chromophores separated by a single chain hydrocarbon bridge (Scheme 2). EPR investigations of the radical anions of α, ω-diarylalkanes, $\mathbf{1}(n)$, and related systems, showed that ET or, more accurately, spin transfer, was rapid on the hyperfine time-scale ($> 10^7$ s^{-1}) for $n = 1$ and 2, but that it was slow for longer chains.[47,48] Indeed, it was the EPR investigations of spin transfer dynamics in radical anions of $\mathbf{1}$ (n) that inspired McConnell to develop his treatment of superexchange. The EPR results for the radical anions of $\mathbf{1}(n)$ were interpreted in terms of a TS mechanism. The rapid spin transfer dynamics observed for the first two members of the series ($n = 1, 2$) were attributed to the short, flexible hydrocarbon chains allowing the two chromophores to adopt the optimal distance and orientation to allow rapid ET to take place directly, TS. The much slower transfer dynamics observed for longer chain lengths was attributed to weak TS interactions, owing to unfavourable hydrocarbon chain conformations, and, by inference, to a negligible superexchange-mediated spin transfer mechanism. In a similar vein, the observation of intra-molecular ET in the series of dinaphthyl radical anions of $\mathbf{2}(n)$, for values of n ranging from 3 to 20 methylene units, was attributed to short-range, TS ET, permitted by the flexible alkyl chain occasionally adopting favourable conformations.[49]

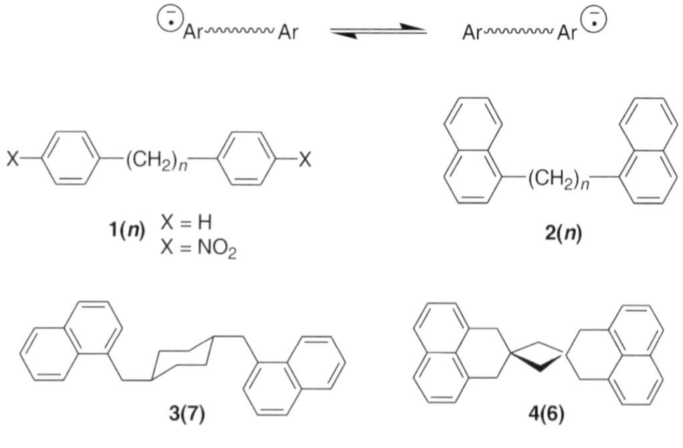

Scheme 2

The belief that TB-mediate ET involving saturated hydrocarbon bridges is generally quite weak, and that it is strongly attenuated with increasing bridge length, prevailed, even into the 1980s. This belief largely derived from taking, too seriously, the numerical estimates presented in McConnell's 1961 paper, notwithstanding that paper's caveat not to do so.[34]

It was this widespread, but misplaced, belief which caused the controversy over the interpretation of the observed rapid intra-molecular ET rates ($\approx 10^7$ s^{-1}) in the radical anion of the semi-flexible *trans*-dinaphthyl-cyclohexane system **3(7)**, where the number in parentheses refers to the number of C–C bonds in one of the relays of the hydrocarbon bridge connecting the pair of chromophores.[50] The observed rapid ET rate was tacitly attributed to a TS mechanism, rather than to a TB mechanism, even though the two naphthalene rings are about 7 Å apart (edge-to-edge). A TS mechanism was likewise advanced to account for the observed rapid intra-molecular ET occurring in the radical anion of **4(6)**.[51]

In spite of this prevailing "TS" mechanistic view of long-range ET, there were some dissenting voices in the 1970s speaking out in favour of a TB-mediated ET mechanism operating in certain systems, the most notable emanating from the Verhoeven group.[52-55] Photophysical investigations of the series of dyads **5(n)** (Scheme 3) revealed the presence of a CT absorption band, due to ET from the dimethoxybenzene donor to the *N*-alkylpyridinium acceptor, for the first two members of the series, **n** = 1 and 2, but not for the higher members.[56-58] The comparatively intense CT band observed for **5(2)** was attributed to efficient through-bond-mediated CT resulting from the molecule adopting the all-*trans* conformation (see Scheme 3).[55] This is an example of the all-*trans* rule of TB interactions which states that the strength of TB coupling through a bridge increases with increasing number of *trans* conformations of *vicinal* C–C bonds; it is shown schematically in Fig. 11.[36,37]

Charge transfer absorption and fluorescence bands were also observed in **6(3)**, in which the nitrogen atom and the dicyanovinyl (**DCV**) chromophores are separated by two hydrocarbon relays, each three C–C bonds in length. Again, this observation was attributed to TB-mediated CT because the orientation of the participating orbitals of the chromophores is optimal for superexchange coupling with the all-*trans* alignment of the cyclohexane bridge bonds, but not for a TS mechanism.[52,54] A particularly revealing photophysical study, which conclusively demonstrated the operation of TB-mediated ET, concerned the systems **7(5)** and **8(5)** which differ only in the stereochemistry about the ring-fusion bond. Both stereoisomers have a pair of relays, five bonds in length, connecting the sulfur and **DCV** chromophores, but they adopt the *trans–trans–gauche* conformation in **7(5)**, and the *gauche–trans–gauche* conformation in **8(5)**. Both CT absorption and fluorescence bands were observed for **7(5)** but neither was observed for **8(5)**. This finding is readily explained in terms of a TB-mediated CT mechanism operating in **7(5)**, but not in **8(5)** which possesses a less favourable conformation of C–C bonds for TB coupling.[53,54]

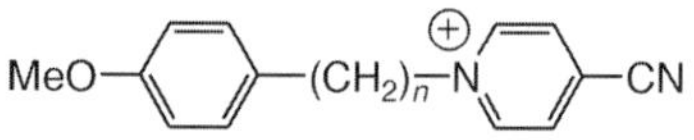

5(n) 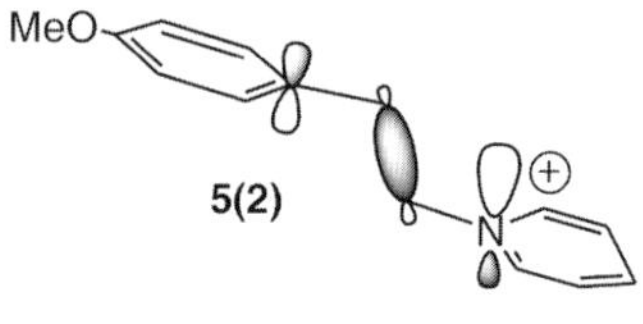   **5(2)**

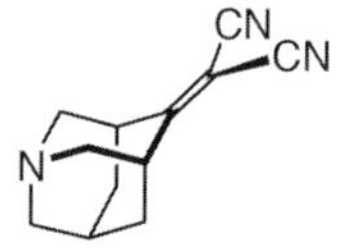

6(3)   **7(5)**   **8(5)**

CT absorption   CT absorption   No CT absorption
CT fluorescence   CT fluorescence   No CT fluorescence

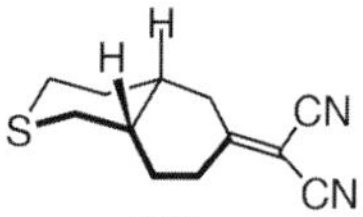

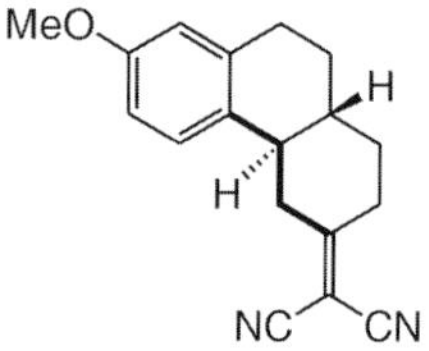

9(3)   **10(5)**

CT fluorescence   CT fluorescence

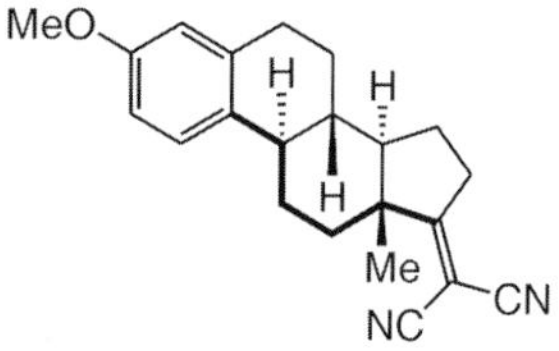

Scheme 3

Photophysical studies on **9(3)** and **10(5)** revealed that extremely rapid photo-induced ET took place in both systems, from the locally excited methoxybenzene donor to the **DCV** acceptor.[59,60] Particularly noteworthy was the observation of CT fluorescence in **10(5)** which was the first documented example of "exciplex-like" emission from a rigid **D–B–A** system with a donor–acceptor separation exceeding three C–C bonds. These data clearly point to TB-mediated ET processes in these molecules.

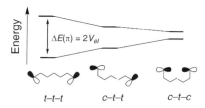

The *all-trans* rule of TB coupling

Fig. 11 The all-*trans* rule of through-bond coupling; the magnitude of $\Delta E(\pi)$ decreases with increasing number of *cisoid* (or *gauche*) conformations in the bridge.

Other investigations of photoinduced ET reactions in various covalently linked organic dyads are providing important mechanistic insights into charge separation processes. In particular, the photophysics of the dyads $11(n)$,[61,62] $12(n)$[63,64] and $13(n)$[65-68] have been extensively studied (Scheme 4). These systems, on account of the conformational flexibility of the polymethylene chains provide valuable information on the inter-play between the dependence of photoinduced ET rates on the conformations of the hydrocarbon bridge and the dynamics of inter-conversion between various conformations of the bridge. The data were interpreted in terms of TS and TB (superexchange) interactions.

In 1980 we reported an interesting observation which implicated a TB-mediated ET mechanism in a chemical reaction. While investigating the dissolving metal reduction (Birch reduction) of a series of benzo-bridge-alkene systems, such as $14(4)$ (Scheme 5), we found that the alkene double bond, but not the aromatic groups, in these systems underwent unusually rapid Birch reduction, which is quite extraordinary because non-conjugated double bonds are generally inert under Birch reduction conditions.[69,70] For example, the double bond in $14(4)$ is reduced some 2900 times more rapidly than the double bond in norbornene. The proposed mechanism involves electron capture by the benzo groups, to give $14a(4)$, followed by reversible (endergonic) intra-molecular ET to the double bond, and subsequent protonation of the resulting radical anion $14b(4)$. Because the complete rigidity of

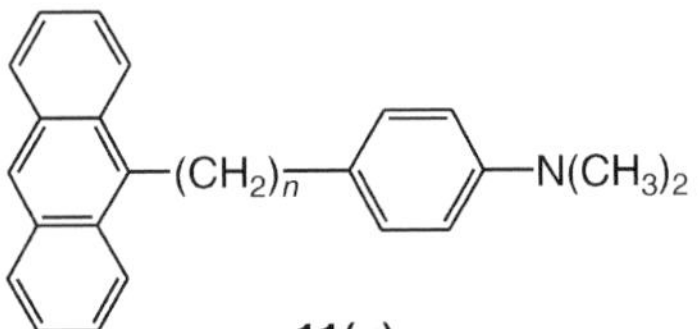

$11(n)$

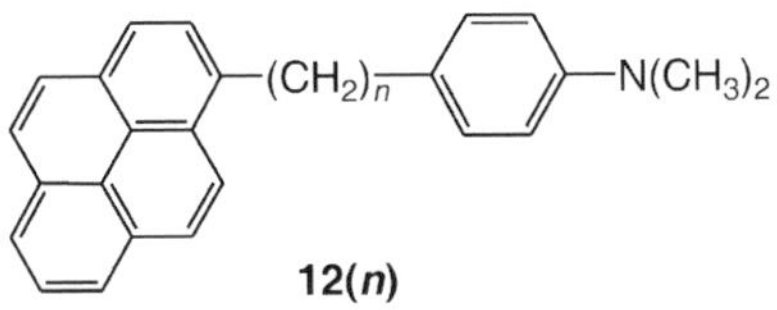

$12(n)$

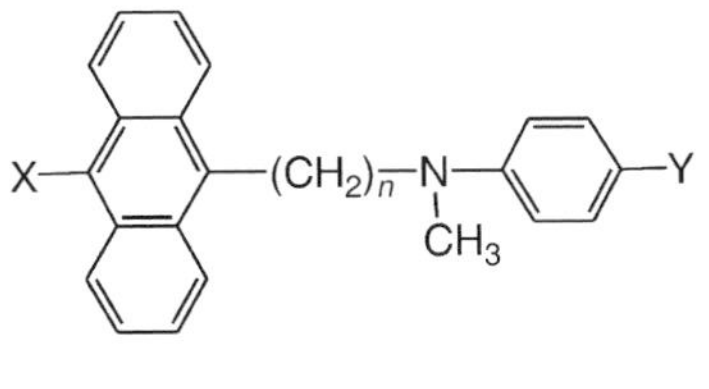

$$13(n) \quad \begin{array}{ll} a & X = CN, \ Y = OMe, \ n = 1 - 4 \\ b & X = CN, \ Y = H, \quad n = 1 - 4 \end{array}$$

Scheme 4

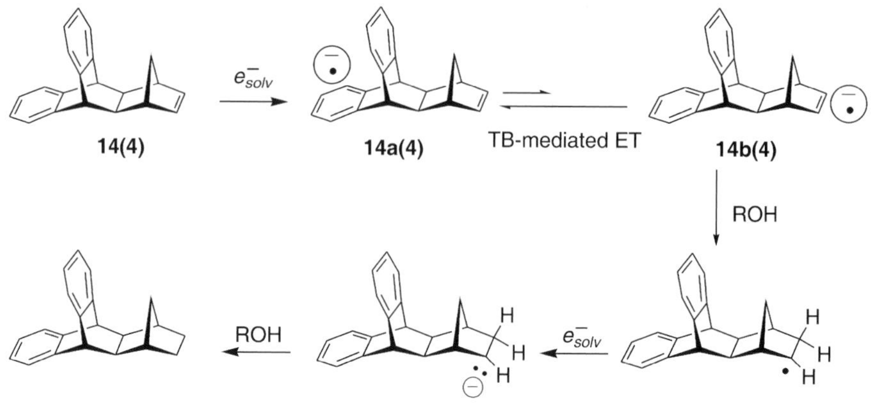

Scheme 5

the norbornylogous bridge in **14(4)** prevents the two chromophores from making close, TS, contact, we proposed that the ET process between **14a(4)** and **14b(4)** occurred by a TB mechanism.[69,70]

To summarise the situation so far, it is fair to say that by the end of the 1970s there was no clear consensus concerning the importance of superexchange-mediated (TB-mediated) electronic coupling through saturated hydrocarbon bridges. To be sure, there were tantalising reports suggesting that this type of superexchange might influence ET dynamics but the thinking of the time was definitely inclined towards the TS electronic coupling mechanism, even for long-range ET processes. However, this inclination was based neither on sound well-established principles, nor on unequivocal experimental data but, rather, on a reluctance to accept that saturated hydrocarbons, with their perceived highly localised and energetically unfavourable σ and σ^* MOs could effectively interact with the π and π^* manifolds of donor and acceptor groups.

The tenability of the TS mechanism as a general mechanism for long-range ET reactions was becoming decidedly less secure with the growing awareness that many biological ET processes in redox proteins were taking place over very large distances (> 20 Å).[1,71–75] While conformational arguments were occasionally mustered in support of the TS-mediated ET mechanism in proteins, which went along the lines that the protein may sometimes adopt a favourable conformation that enables the redox couple to lie in close spatial proximity to each other, they lacked cogency. More disturbing was the revelation in the 1980s that the primary ET events in the photosynthetic reaction centres of certain photosynthetic bacteria[10] took place with extraordinary rapidity (subnanosecond timescale) over distances exceeding 15 Å.[76–78]

The emerging investigations of biological ET made the need to obtain an unambiguous distance dependence of ET rates mediated by saturated hydrocarbon bridges even more urgent, not only because of the need to calibrate the accuracy of the McConnell model, but also because saturated hydrocarbon chains, acting as simplified models for proteins, should provide crucial mechanistic insights into biological ET reactions.

Accordingly, in the early 1980s, the problem of the distance dependence of ET rates became one of the outstanding challenges in the entire ET field. We solved this problem in two steps, firstly, by using photoelectron spectroscopic (PES) and electron transmission spectroscopic (ETS) techniques and secondly, by making direct measurements of ET rates. The key to our success was the design and synthesis of an elegant series of **D–B–A** systems.

ORBITAL INTERACTIONS IN POLYNORBORNANE-DIENES

Our first step towards determining, unequivocally, the strength and the distance dependence of TB-mediated electronic coupling within saturated hydrocarbon bridges stemmed from our PES and ETS studies on the first three members of the series of totally rigid, symmetrical, polynorbornane-dienes **15(4)**–**15(6)**, where the numbers in parentheses refer to the numbers of C–C bridge bonds connecting the two double bonds (Fig. 12). The experimental studies were supplemented with

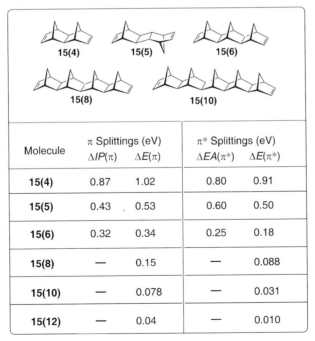

Molecule	π Splittings (eV)		π^* Splittings (eV)	
	$\Delta IP(\pi)$	$\Delta E(\pi)$	$\Delta EA(\pi^*)$	$\Delta E(\pi^*)$
15(4)	0.87	1.02	0.80	0.91
15(5)	0.43	0.53	0.60	0.50
15(6)	0.32	0.34	0.25	0.18
15(8)	—	0.15	—	0.088
15(10)	—	0.078	—	0.031
15(12)	—	0.04	—	0.010

Fig. 12 Koopmans' theorem splitting energies for **12(n)**, obtained from ionisation potentials, $\Delta IP(\pi)$, and electron affinities, $\Delta EA(\pi^*)$, together with HF/3-21G Koopmans' theorem splittings, $\Delta E(\pi)$, and $\Delta E(\pi^*)$.

simple *ab initio* MO calculations which enabled us to investigate orbital interactions in longer members of the series, **15**(*n*; *n* = 8–12), which were not synthetically accessible. Fig. 12 summarises the experimental gas phase π-ionisation potential splittings, $\Delta IP(\pi)$, and π^*-electron affinity splittings, $\Delta EA(\pi^*)$ (from PES and ETS measurements, respectively) for **15(4)–15(6)**.[37,79–81] Both $\Delta IP(\pi)$ and $\Delta EA(\pi^*)$ for these dienes were found to decay exponentially with increasing number, *n*, of bridge σ-bonds according to equations (11) and (12), respectively:

$$\Delta IP(\pi) = 6.0 \exp(-0.50n); \qquad \beta_{el}(\pi) = 1.0 \text{ bond}^{-1} \tag{11}$$

$$\Delta EA(\pi^*) = 9.0 \exp(-0.58n); \qquad \beta_{el}(\pi^*) = 1.2 \text{ bond}^{-1} \tag{12}$$

These results, obtained between 1980 and 1983, were significant because they gave, for the first time, an unequivocal estimate of the distance dependence of TB-mediated electronic coupling through a saturated hydrocarbon bridge. Our interpretation of the results was unequivocal for the following three reasons. Firstly, the complete rigidity of our novel bridges precisely fixes the distance between the double bonds. Secondly, the molecular symmetry (C_{2v}) of the dienes allowed us to equate the observed splitting energies with twice the magnitude of the electronic coupling between the π MOs, and between the π^* MOs, of the double bonds. Thirdly, the measurements were carried out in the gas phase, thereby avoiding any complicating solvent effects on the strength of the electronic coupling.

The relevance of our PES and ETS studies to HT and ET processes is that the PES $\Delta IP(\pi)$ splitting energies for **15(4)–15(6)** are, to a good approximation, equal to twice the respective electronic coupling terms, V_{el}, for hole transfer in the diene radical cations (Scheme 6a). Likewise, the ETS $\Delta EA(\pi^*)$ values for these dienes are approximately equal to twice the respective electronic coupling terms, V_{el}, for ET in the diene radical anions (Scheme 6b). The derivation of these relationships is straightforward and has been explained in detail elsewhere.[14,38,41] A simple MO description of TB coupling in these and other systems has also been thoroughly described in the literature.[14,36,38,41,79]

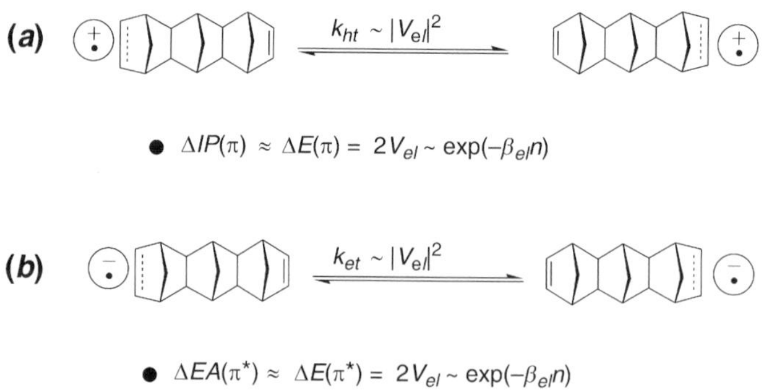

(a) $\quad k_{ht} \sim |V_{el}|^2$

● $\Delta IP(\pi) \approx \Delta E(\pi) = 2V_{el} \sim \exp(-\beta_{el}n)$

(b) $\quad k_{et} \sim |V_{el}|^2$

● $\Delta EA(\pi^*) \approx \Delta E(\pi^*) = 2V_{el} \sim \exp(-\beta_{el}n)$

Scheme 6

The π, π and π^*, π^* splitting energies were calculated using Hartree–Fock *ab initio* SCF MO theory.[14,38,39,79,80,82–85] These calculations were carried out within the context of Koopmans' theorem (KT),[86] which equates the ionisation potential and the electron affinity to the negative of the energy of the orbital that is associated with the ionisation process. The KT splitting energies, $\Delta E(\pi)$ and $\Delta E(\pi^*)$, for **15(4)–15(12)**, obtained using the 3-21G basis set, are listed in Fig. 12. The agreement between the computed and experimental data for **15(4)–15(6)** is quite good, thereby showing that KT calculations can provide quantitatively reliable estimates of the splitting energies for higher members of the series. Fitting the KT/3-21G splitting energies to an exponential decay gives the following β_{el} values ($r^2 = 0.988$ in both cases):

$$\Delta E(\pi) = 4.0 \exp(-0.39n); \qquad \beta_{el}(\pi) = 0.78 \text{ bond}^{-1} \tag{13}$$

$$\Delta E(\pi^*) = 7.0 \exp(-0.55n); \qquad \beta_{el}(\pi^*) = 1.1 \text{ bond}^{-1} \tag{14}$$

There is overall fairly good agreement between the computed and experimental damping factors (cf. equations (11)–(14)), and the average values from these two methods are: $\beta_{el}(\pi) = 0.89 \text{ bond}^{-1}$ and $\beta_{el}(\pi*) = 1.15 \text{ bond}^{-1}$.

Three important conclusions were drawn from our studies of the series of dienes **15(n)**. Firstly, the π, π and π^*, π^* splitting energies are caused by TB interactions and not by TS interactions between the two double bonds. Thus, the calculated TS splitting energies between two isolated ethene molecules, having the same spatial relationships as the double bonds in **15(n)**, were found to be negligible compared to those calculated for **15(n)**.[14,87]

The second conclusion concerns the magnitudes of the TB splitting energies for **15(n)** – they are huge! – at least, in comparison with the typical value of about 3 meV (25 cm^{-1}), for the electronic coupling associated with non-adiabatic ET. For example, in the case the 12-bond system, **15(12)**, the V_{el} values for HT and ET, derived from $\Delta E(\pi)$ and $\Delta E(\pi^*)$, are 20 meV (≈ 160 cm^{-1}, or 2 kJ/mol) and 5 meV (≈ 40 cm^{-1}, or 0.5 kJ/mol), respectively, notwithstanding the 13.5 Å edge-to-edge separation – as the nano-crow flies! – between the two double bonds. Indeed, for this separation, the TS interaction between the double bonds is estimated to be $< 10^{-10}$ eV. Noting that a value of about 3 meV (25 cm^{-1}) for V_{el} is sufficient for promoting rapid ET,[4] the predicted V_{el} values of 20 and 5 meV for HT and ET, respectively, in **15(12)** suggest that TB-mediated HT and ET should be extremely rapid over very large distances.

The third, and possibly the most important conclusion is that the distance dependence of TB-mediated electronic coupling in **15(n)** is much weaker than that predicted by the simple McConnell model. In that model, a β_{el} value of about 2.3 bond^{-1} is estimated, whereas the average β_{el} value of *ca.* 1.0 bond^{-1} for **15(n)** is 2.3 times smaller! Thus, the attenuation in the ET rate, mediated by superexchange through a saturated hydrocarbon bridge, could be as small as *ca.* 2.5 bond^{-1}, as opposed to *ca.* 10 bond^{-1}, predicted from the simple McConnell model.

Parenthetically, the reason why the simple McConnell model greatly over-estimates the distance dependence behaviour for TB-mediated electronic coupling is the model's neglect of non-nearest neighbour interactions within the bridge, which, in fact, are quite significant. This may be seen from the HF/3-21G matrix elements for interactions between localised natural bond orbitals (NBOs)[88] in an alkane bridge (Fig. 13).[84,89] Although the matrix element between two adjacent C–C NBOs, t, has a large value of -4.1 eV, those intra-bridge couplings which skip over one bond (t') and two bonds (t'') are by no means negligible, having values of 0.80 and -0.29 eV, respectively. In fact, inclusion of both the t' and t'' matrix elements in a non-perturbative McConnell treatment of alkane dienes leads to a β_{el} value in good agreement with that obtained from a full Hartree–Fock calculation.[89]

By 1983, it had been established that TB-mediated electronic coupling through saturated hydrocarbon bridges is significant and that it had major implications for the understanding of long-range ET processes, not only taking place through saturated hydrocarbon moieties, but also through other saturated, or nearly saturated systems, such as proteins. My 1982 *Accounts of Chemical Research article* concluded with the following remark:

> Many other electron-transfer reactions may be influenced by through-bond effects. Indeed the possibility of either an electron or a "positive hole" being "transmitted" through bonds, over large distances within a molecule, is intriguing and is being actively investigated by ourselves and, I hope, others!"[37]

This intriguing possibility was, indeed, being actively investigated, and this leads to the second step in our adventures.

THE DISTANCE DEPENDENCE OF LONG-RANGE ET RATES

The year 1983 got off to a brisk start with the appearance of a communication by Calcaterra and co-workers, in which they reported observing very rapid thermal intra-molecular ET in radical anions of **D–B–A** dyads, such as **16(9)** and **17(10)**, in which the chromophores are attached to a rigid steroid bridge (Scheme 7).[90] The edge-to-edge distance between the two chromophores in these molecules is about 11 Å and they are separated by nine C–C bonds. The radical anions of **16(9)** and **17(9)** were generated by pulse radiolysis in 2-methyltetrahydrofuran at room

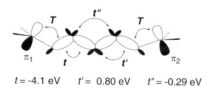

$$t = -4.1 \text{ eV} \qquad t' = 0.80 \text{ eV} \qquad t'' = -0.29 \text{ eV}$$

Fig. 13 HF/3-21G computed matrix elements for interactions between various natural bond orbitals (NBOs).

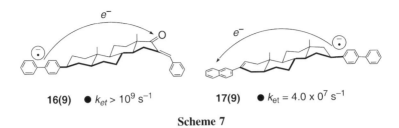

16(9) ● $k_{et} > 10^9$ s^{-1} **17(9)** ● $k_{et} = 4.0 \times 0^7$ s^{-1}

Scheme 7

temperature and the ET rates were determined from the rates at which the radical anions approached equilibrium. The finding of extremely rapid ET rates, subnanosecond in the case of **16(9)**, over such large inter-chromophore distances surprised the authors "in view of the insulating nature of the σ-bonded framework";[90] and they concluded that their observed ET reactions "must be described as long-range electron tunnelling",[90] which is equivalent to saying that the ET processes were taking place by a superexchange mechanism involving the steroid bridge.

These findings by Calcaterra and co-workers, were, of course, entirely consistent with our findings of large TB interactions in the norbornylogous diene series **15(n)**.[37] What was needed, however, was a direct, unambiguous determination of the distance dependence of ET rates through a rigid hydrocarbon bridge of well-defined but variable length. Our successful demonstration that the norbornylogous bridge is an ideal entity for studying TB interactions placed us in a strong position to solve this distance dependence problem, and the breakthrough came with our design and synthesis, over the years, of the novel series of bichromophoric systems **18(n)** (Scheme 8).[91–96]

These dyads, like the series **15(n)**, are based on the rigid norbornylogous bridge whose length ranges from four bonds (4.6 Å, edge-to-edge) to 13 bonds (14.2 Å). The dimethoxynaphthalene (**DMN**) donor and the **DCV** acceptor groups are well-suited for studying both photoinduced charge separation in the neutral systems, and thermal charge-shift ET in the radical anions. The molecules **18(n)** look somewhat spartan in appearance but they possess great elegance and beauty, and they are ideal for the task in hand. They are rigid and they have symmetry. They are free of unnecessary functionality which might cloud the interpretation of experimental data. The variation in the length of the norbornylogous bridge in **18(n)**, ranging from four bonds to thirteen, is large enough to obtain a precise distance dependence of ET rates.

Many ET studies have been carried out on **18(n)**, and their cognates, most of which have been in fruitful and rewarding collaborations with two great Dutch groups, namely the Verhoeven and Warman research teams. Only the more important results will be presented here, as more comprehensive accounts of our work have been given elsewhere.[41,55,97–99] The following two points should be borne in mind in the following discussion. First, β values will, in general, be discussed in units of bond^{-1}, rather than in Å^{-1}, because it seems more logical to do so when addressing the distance dependence of electronic coupling through

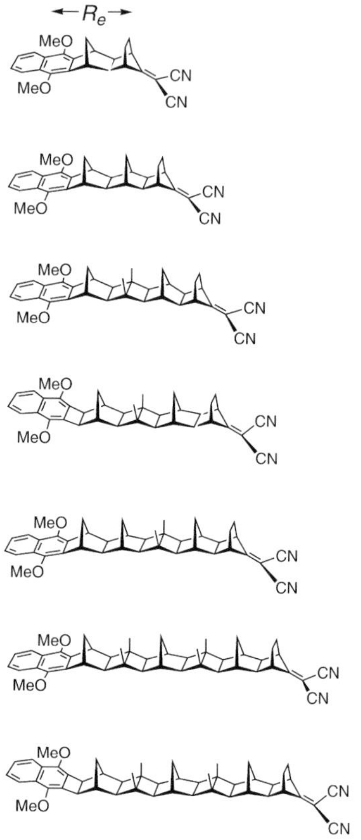

$$\overleftarrow{R_e}\overrightarrow{} \qquad\qquad R_e \text{ (Å)}$$

18(4)   4.6

18(6)   6.8

18(8)   9.4

18(9)   10.9

18(10)   11.5

18(12)   13.5

18(13)   14.2

Scheme 8

bonds, rather than TS. The two units are related however, by dividing the number of bonds in a particular bridge by the edge-to-edge distance between the points of attachment of the chromophores to the bridge; this reduces to the following equivalence: one C–C bond $\equiv 1.11$ Å for saturated hydrocarbon bridges possessing the all-*trans* (i.e., all-*anti*) conformation.[100,101] Secondly, as mentioned earlier, experimentally determined damping factors derived from ET rates, and which have not been corrected for the distance dependence of Franck–Condon factors, are denoted by β, whereas those damping factors which refer to the distance dependence of V_{el} are denoted by β_{el}.

The first studies carried out on **18(n)** were measurements of the rates of photoinduced intra-molecular ET (charge separation)[92,102] and subsequent charge recombination (CR).[99,101,103,104] Photoinduced charge separation in **18(n)**, illustrated in Fig. 14, involves initial formation of the locally excited singlet state of the **DMN** chromophore, followed by charge separation, to form the singlet CS state $^+$**DMN[n]DCV** $^-$. The photoinduced charge separation rate data, summarised in

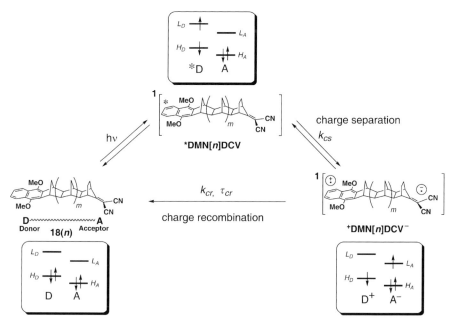

Fig. 14 A schematic of the photoinduced charge separation and charge recombination processes in **18(n)**. A simple orbital diagram is provided which captures the essentials of the ET processes. H_D, donor HOMO; L_D, donor LUMO; H_A, acceptor HOMO; L_A, acceptor LUMO. Note that all depicted processes are assumed to take place on the singlet multiplicity manifold.

Fig. 15, were initially obtained from the **DMN** fluorescence lifetime data,[92,102] but they have recently been measured using subpicosecond time-resolved transient absorption spectroscopy (pump–probe).[105]

The intra-molecular photoinduced charge separation rates for **18(n)** are amazingly rapid; even for the 13-bond dyad **18(13)**, in which the **DMN** and **DCV** groups are 14 Å apart, the ET rate is a massive 10^8 s^{-1}. The ET rates for these dyads are fairly insensitive to solvent polarity, provided the charge separation process remains exergonic.[92] For example, the photoinduced charge separation rate for **18(8)** is only 1.5 times faster in acetonitrile (dielectric constant, $\epsilon_s = 37.5$), than in cyclohexane ($\epsilon_s = 2.02$). This somewhat surprising result is due to the fact that photoinduced charge separation in **18(n)** is virtually activationless; i.e., $-\Delta G^0 \approx \lambda$; see equation (2). Changing the polarity of the solvent causes compensating changes in the values of ΔG^0 and λ_s, with the result that $\Delta G\ddagger$ remains close to zero.[92,106]

It seems inconceivable that such rapid ET rates observed in **18(n)** could be due to a TS effect, particularly in light of our earlier studies of **15(n)** (Fig. 12), which proved the existence of strong through-bond coupling involving the norbornylogous bridge. Nevertheless, we obtained conclusive proof that the observed photoinduced ET reactions in **18(n)** are due to the superexchange mechanism by taking advantage

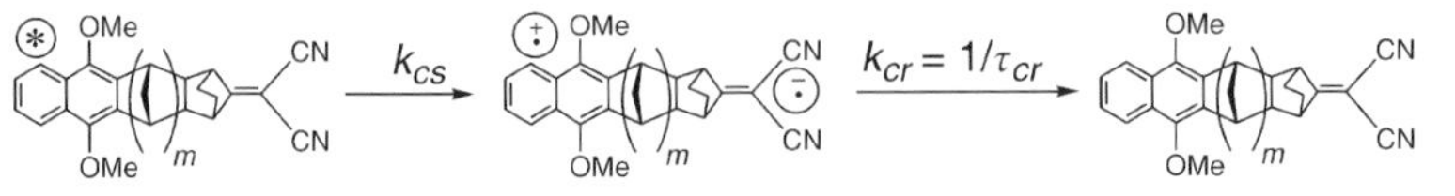

	$k_{cs}/10^8$ s^{-1}		τ_{cr} (ns)
	fluorescence measurements	pump-probe measurements	
18(4)	>> 5000	28000	–
18(6)	3000	4890	0.5
18(8)	670	1160	2.5
18(9)	250	647	—
18(10)	120	180	43
18(12)	13	—	297
18(13)	1.2	—	1050

Fig. 15 Rate data for photoinduced charge separation and subsequent charge recombination in the dyads **18(n)**. Charge separation rates, k_{cs}, in THF at 20°C were determined both from fluorescence lifetimes[92,102] and by pump-probe (time-resolved transient absorption) spectroscopic measurements.[105] The mean lifetimes towards charge recombination, τ_{cr}, were obtained from time-resolved conductivity measurements in 1,4-dioxane.[99,101,103,104]

of the all-*trans* rule of TB coupling (Fig. 11). If TB coupling were mediating the observed ET reactions in **18(n)**, then the ET rate should be modulated by changing the configuration of the norbornylogous bridge, in accordance with the all-*trans* rule. This was tested using "kinked" dyads, such as **19(8)** which possesses two *cisoid* (or *gauche*) arrangements of *vicinal* C–C bridge bonds in each of the two 8-bond relays (Fig. 16). The photoinduced ET rate for the all-*trans* system **18(8)** was found to be an order of magnitude faster than that for the "kinked" molecule **19(8)** (Fig. 16) – in spite of the fact that the chromophores are 0.4 Å *further apart* in the former system – thereby elegantly confirming the operation of TB-mediated ET in **18(n)**.[107,108] The all-*trans* rule has been used to probe the involvement of superexchange in ET processes in other systems.[109]

Reasonable exponential fits were obtained for the distance dependence of the photoinduced charge separation rates (in THF) for **18(n)**, obtained from both fluorescence[92] and pump-probe measurements,[105] and they are shown in Fig. 17. The phenomenological damping factors derived from the two sets of measurements are:

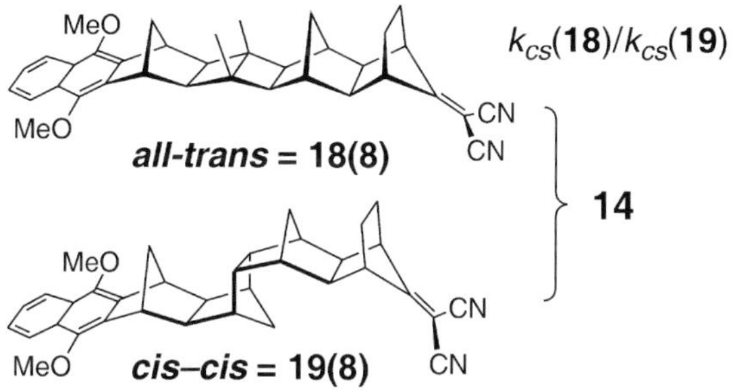

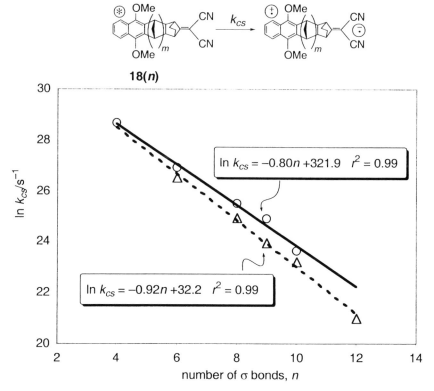

Fig. 16 Ratio of the photoinduced charge separation rates for **18(8)** and **19(8)**, measured in acetonitrile.[107,108] The two bridge relays connecting the two chromophores in the dyad **18(8)** have an all-*trans* arrangement of C–C bonds, whereas in **19(8)**, two pairs of *vicinal* C–C bonds have a *cis* (or *gauche*) arrangement.

Fig. 17 Plots of $\ln k_{cs}$ versus the bridge length, expressed as the number, n, of σ bonds, for **18(n; n = 4, 6, 8, 9, 10, 12, 13)** in THF solvent, determined from fluorescence lifetimes ($\triangle$) and pump-probe measurements ($\bigcirc$).

$$\beta(\text{fluorescence}) = 0.92 \text{ bond}^{-1}(\text{or } 0.82 \text{ Å}^{-1}) \tag{15a}$$

$$\beta(\text{pump-probe}) = 0.80 \text{ bond}^{-1}(\text{or } 0.71 \text{ Å}^{-1}) \tag{15b}$$

The initial studies of **18(*n*)**[92,102] provided the first definitive experimental evidence for the exponential distance dependence behaviour of ET rates for any long-range ET process. Of greater significance, our data demonstrate, quite unequivocally, that the magnitude of superexchange coupling through a saturated hydrocarbon bridge is not only surprisingly strong, giving rise to subnanosecond ET lifetimes for donor–acceptor distances exceeding 12 Å, but it also displays a remarkably weak distance dependence, with a β value of about 0.80–0.9 bond^{-1}.

The close agreement between β, for photoinduced charge separation in **18(*n*)**, and β_{el}, determined from PES, ETS and computational studies on **15(*n*)**, is gratifying, and further strengthened our assertion that the extremely rapid rates of photoinduced ET observed in **18(*n*)** are the result of a superexchange mechanism mediated by the norbornylogous bridge.

Within the context of a simple orbital representation, the active MOs involved in photoinduced charge separation in **18(*n*)** are the **DMN** donor LUMO (L_D) and the **DCV** acceptor LUMO (L_A) (Fig. 14), and the interaction of these MOs with the bridge orbitals determines the magnitude of V_{el} for this ET process and, hence, the magnitude of β_{el} (and β, to a lesser extent). Our systems offer the opportunity for studying several other types of ET processes, involving different active orbitals and, hence, to investigate the sensitivity of the magnitude of β to the different types of active orbitals. Some examples of these studies are now presented.

Photoinduced ET in **18(*n*)** leads to the formation of the singlet CS state, $^{+}$**DMN[*n*]DCV**$^{-}$. The active MOs involved in the direct charge recombination process, from this CS state to the ground state, are H_D and L_A (Fig. 14). The lifetimes, τ_{cr}, of the CS states, obtained from time-resolved microwave conductivity (TRMC) measurements in 1,4-dioxane, are given in Fig. 15.[99,101,103,104] As expected, the lifetimes of these CS states increase exponentially with increasing bridge length due, in large part, to the decaying strength of V_{el} for CR. The β value for CR is:

$$\beta(\text{CR}) = 1.1 \text{ bond}^{-1}(\text{or } 0.98 \text{ Å}^{-1}) \tag{16}$$

Fluorescence spectra of the dyads **18(*n*)** in butyl ether revealed the presence of discrete CT emission bands for bridge lengths up to 10 sigma bonds.[110] This was a significant observation because these CT fluorescence bands enabled us to calculate the electronic coupling element, V_{el} for the CR process and, hence, to obtain the associated β_{el} value – as distinct from the phenomenological β value of equation (16), determined from CR lifetime measurements. The results, presented in Fig. 18a, give a β_{el} value of:

$$\beta_{el}(\text{CR}) = 1.0 \text{ bond}^{-1}(\text{or } 0.89 \text{ Å}^{-1}) \tag{17}$$

The almost identical values of β and β_{el} for CR in $^{+}$**DMN[*n*]DCV**$^{-}$ indicate that the distance dependence of the CR lifetimes is determined mainly by the distance

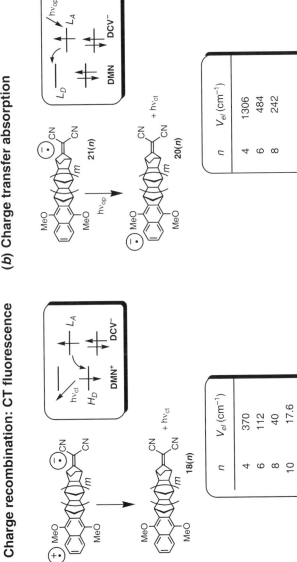

Fig. 18 (a) Values of the electronic coupling matrix element, V_{el}, calculated from the charge transfer fluorescence data for **18(n)** in butyl ether.[110] The CT fluorescence arises from the charge-separated state. (b) Values of V_{el} calculated from the optical electron transfer bands for the radical anions **21(n)**, in which the unpaired electron is localised on the **DCV** moiety, to the radical anion **20(n)**, in which the unpaired electron is localised on the **DMN** group.

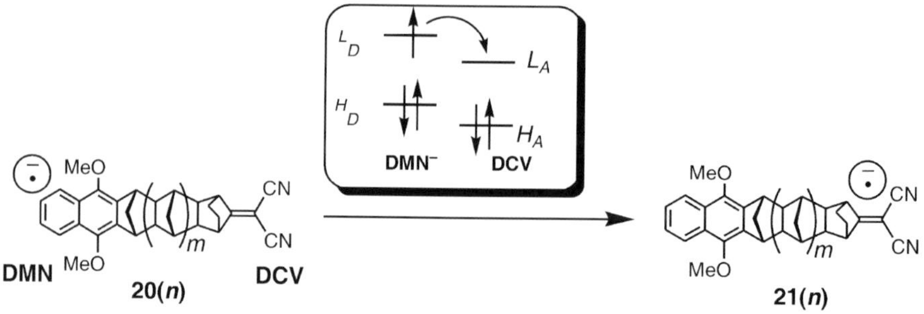

Scheme 9

dependence of V_{el}, and that the distance dependence of the Franck–Condon factors is small.

Thermal charge-shift intra-molecular ET in the radical anions **20(n)**, involving L_D and L_A as the active MOs (Scheme 9), was investigated using pulse radiolysis,[111] but the rates were found to exceed the time resolution of the equipment ($> 10^9$ s^{-1}) for all radical anions, including **20(12)**.[111] However, optical ET bands were observed in the visible-near-infrared absorption spectra of the product radical anions, **21(n)**. These bands arise from optical ET, from the **DCV** radical anion group to the **DMN** moiety. Analysis of these bands using Hush theory[112,113] led to estimates of V_{el} for the charge-shift reaction (Fig. 18b). These data give a β_{el} value of:

$$\beta_{el}(\text{charge-shift}) = 0.84 \text{ bond}^{-1}(\text{or } 0.75 \text{ Å}^{-1}) \qquad (18)$$

The final example of the distance dependence of ET dynamics in norbornylogous bridged systems is photoinduced charge-shift ET in **22(4)** and **22(6)** which possess the *N*-methylpyridinium acceptor (Scheme 10).[114] Local excitation of the **DMN** donor in the dyads results in charge-shift, from L_D to L_A, to form the distonic radical

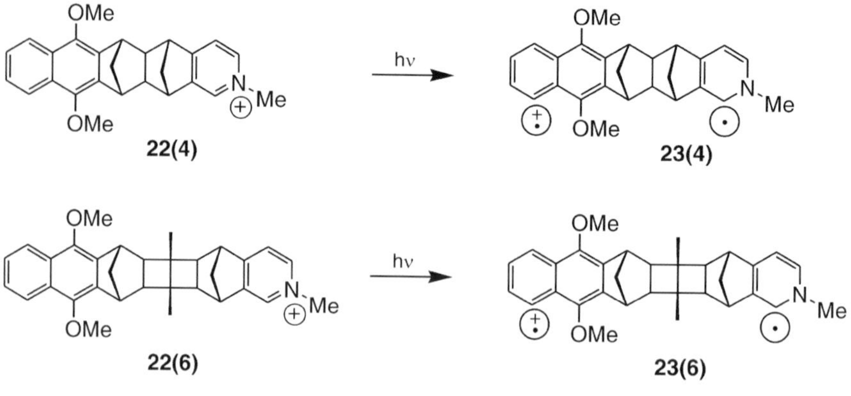

Scheme 10

cations $23(n; n = 4, 6)$. From these limited data, the following approximate β value is obtained:

$$\beta_{el}(\text{charge-shift}) = 0.88 \text{ bond}^{-1} (\text{or } 0.77 \text{ Å}^{-1}) \tag{19}$$

Summarising the results for $18(n)$–$22(n)$, the experimentally determined β and β_{el} values for several different types of ET reactions – photoinduced charge separation, charge recombination, thermal charge-shift, photoinduced charge-shift, and optical ET – having different driving forces, and occurring in both the Marcus normal and inverted regions, all lie within the rather narrow range of $\beta = 0.84$–1.1 bond^{-1} (equations (15)–(19)), with an average value:

$$\beta(\text{average}) = 0.92 \text{ bond}^{-1} \tag{20}$$

The near constancy of β in these studies strongly indicates that its value reflects the distance dependence of TB coupling through the norbornylogous bridge. The insensitivity of β (and β_{el}) to the type of ET process may be explained as the consequence of the very large energy gap, Δ, between the π (and π^*) levels of the chromophores and the bridge σ (and σ^*) levels. Thus, within the context of the McConnell model, the variation of β with the energy gap Δ is given by equation (21) which is derived from equation (10), assuming constancy of the intra-bridge coupling interactions, t.

$$\left(\frac{\delta\beta}{\delta\Delta}\right)_t = \frac{2}{\Delta} \tag{21}$$

Thus, the sensitivity of β to changes in the magnitude of $|\Delta|$ varies inversely with the magnitude of $|\Delta|$, and so it isn't surprising that β doesn't vary too much with the different types of chromophore orbitals involved in the various ET reactions studied in $18(n)$–$22(n)$.

Equation (21) also leads us to predict that, barring significant changes in the magnitudes of the various intra-bridge coupling interactions, t, t', etc. (Fig. 13), β shouldn't vary much with the type of saturated hydrocarbon bridge. Experimental results for a representative set of systems possessing different saturated hydrocarbon bridges, shown in Fig. 19, do, indeed, appear to support this prediction.

Thus, β for optical (intervalence) ET, mediated by a rigid oligospirocyclobutane bridge is 0.88 bond^{-1} (Fig. 19a).[115] Measurements of the distance dependence of both thermal intra-molecular electron and hole transfer in radical anions and radical cations of rigidly fused cyclohexane and steroid bridges, generated by pulse radiolysis, all gave β values of approximately 0.90 bond^{-1} (Fig. 19b).[116–118] The distance dependence of ET rates through flexible single chain polymethylene bridges has also been investigated. In two studies, (Fig. 19c and d), the bridge was forced to adopt the all-*trans* conformation through the encapsulation of the polymethylene chain by β-cyclodextrin. The β values obtained from photoinduced ET rate measurements using the $\mathbf{Ru(II)}$–$\mathbf{(CH_2)_n}$–$\mathbf{MV^{2+}}$ dyads[119] (Fig. 19c) and the $\mathbf{naphthalene}$–$\mathbf{(CH_2)_n}$–$\mathbf{MV^{2+}}$ dyads[120] (Fig. 19d) are 1.18 and 1.09 bond^{-1},

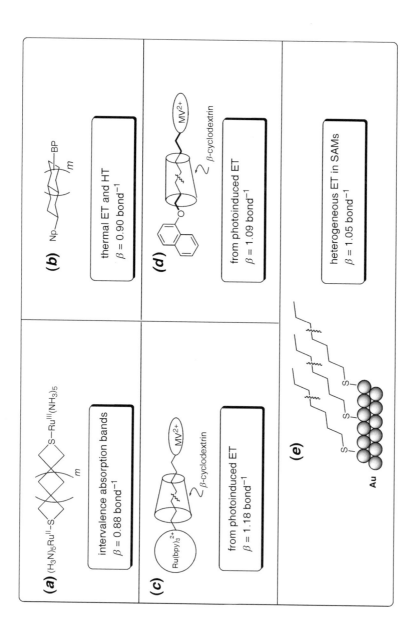

(a) $(H_3N)_5Ru^{II}$–S $\diamond\diamond\diamond$ S–$Ru^{III}(NH_3)_5$

intervalence absorption bands
$\beta = 0.88$ bond^{-1}

(b) Np $\diamond\diamond$ BP

thermal ET and HT
$\beta = 0.90$ bond^{-1}

(c) Ru(bpy)$_3^{2+}$ ... MV^{2+}
β-cyclodextrin

from photoinduced ET
$\beta = 1.18$ bond^{-1}

(d) ... MV^{2+}
β-cyclodextrin

from photoinduced ET
$\beta = 1.09$ bond^{-1}

(e) Au

heterogeneous ET in SAMs
$\beta = 1.05$ bond^{-1}

respectively. A recent determination of the distance dependence of electron tunnelling through self-assembled monolayers (SAMs) comprising polymethylene chains (Fig. 19e), using conducting probe atomic force microscopy, yielded a β_{el} value of 1.05 bond^{-1}.[121]

From these extensive studies, we may conclude that the distance dependence of ET rates, mediated by a superexchange mechanism involving saturated hydrocarbon bridges, is largely independent of the nature of the hydrocarbon and has an average β value of 0.99 ± 0.2 bond^{-1}.

5 A more detailed analysis at TB coupling

A QUESTION OF INTERFERENCE

The prediction of the constancy of β_{el} for different saturated hydrocarbon bridges must be treated with caution because it was derived using the McConnell nearest neighbour model of superexchange. Could the presence of non-nearest neighbour interactions give rise to different β_{el} values for different saturated hydrocarbon bridges? In particular, if two chromophores are connected to each other by several TB-coupling relays, could interactions between the relays affect both the strength and distance dependence of the TB-induced coupling between the chromophores? We asked these questions in 1990 and proceeded to investigate them using *ab initio* MO calculations (HF/3-21G) within the context of Koopmans' theorem (KT). In the following discussion, we focus on hole transfer processes in radical cations of dienes; the KT calculations are, therefore, restricted to π, π MO splitting energies, $\Delta E(\pi)$.

Fig. 20 summarises the limiting β_{el} values derived from the $\Delta E(\pi)$ values for four diene systems, **24(n)–26(n)** and **15(n)**, differing in the type of hydrocarbon bridge. The distance dependence of $\Delta E(\pi)$ is not precisely monoexponential over the whole range of bridge lengths studied, although it does tend to become so for large distances (i.e., $n > 10$ bonds), and the limiting β_{el} values are obtained at this asymptotic limit. Thus, the limiting β_{el} value of 0.68 bond^{-1} for **15(n)** is slightly smaller than the previously cited value of 0.78 bond^{-1} (equation (13), derived from

---

Fig. 19 Some β values for dyads possessing different saturated hydrocarbon bridges. (a) β for an oligospirocyclobutane bridge, determined from the analysis of intervalence bands.[115] (b) β for fused cyclohexane and steroid bridges, obtained form thermal charge-shift (both electron transfer and hole transfer) studies in the respective radical anions and radical cations.[116–118] (c) and (d) β for a single chain alkane bridge, maintained in the all-*trans* conformation by a β-cyclodextrin host molecule, determined by photoinduced ET, using either (c) a Ru(II) donor,[119] or (d) a naphthalene donor.[120] (e) β for a single chain alkane bridge in a self-assembled monolayer, determined using conducting probe atomic force microscopy.[121]

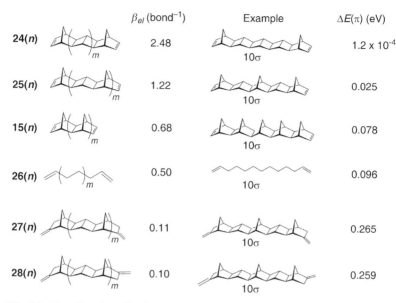

Fig. 20 Limiting β_{el} values for the π, π splitting energies, $\Delta E(\pi)$ for some dienes. The β_{el} values were obtained by fitting the HF/3-21G $\Delta E(\pi)$ values for consecutive members of the series to $\Delta E(\pi) = A \exp(-0.5\beta_{el}n)$, where n is the number of bonds in a relay connecting the two double bonds, a relay being defined as the shortest path connecting the two double bonds. The $\Delta E(\pi)$ value for **24(10)** was calculated from extrapolation of that for **24(8)** and using a β_{el} value of 2.48 bond^{-1}.

shorter members of the series). The π, π splittings calculated for the 10-bond dienes, **24(10)**–**26(10)** and **15(10)**, range from 1.2×10^{-4}eV to 0.096 eV, and the β_{el} values range from 0.50 to 2.48 bond^{-1}. The largest splitting energies and the smallest β_{el} value are associated with the simple all-*trans* alkane bridge diene, **26(n)**, and the smallest splitting energies and the largest β_{el} value are associated with **24(n)** in which the bridge consists of linearly fused cyclobutane rings, capped at each end by a norbornene group. These results surprised us because simple perturbation theory predicts that the electronic coupling through a bridge possessing m non-interacting relays of identical lengths – a relay being defined as a single chain of C–C bonds connecting the two chromophores – should be m-times the strength of the coupling through one of the relays, while the β_{el} value should be the same in both cases.[122,123]

That this prediction fails for **15(n)**, **24(n)**, **25(n)**, and other dienes[41,123] possessing multiple TB relays suggests that the relays are interacting with each other, but in a destructive manner which degrades the magnitude and the distance dependence of the TB-mediated electronic coupling. A detailed NBO analysis of **15(n)**, **24(n)**, and **25(n)** confirmed this suspicion. The NBO analysis may be neatly and concisely explained using the simple McConnell nearest-neighbour superexchange model. If we relax the condition that the magnitude of the coupling, t, between adjacent localised bridge orbitals is the same throughout the bridge, then the more general form of the McConnell equation becomes:

$$\Delta E(\pi) = \frac{-2T_{D,1}T_{n,A}}{\Delta_1} \prod_{i=1}^{n-1} \frac{t_{i,i+1}}{\Delta_{i+1}} \tag{22}$$

where n is the number of localised bridge orbitals in a particular relay; φ_i, is the localised orbital of the ith C–C bond in the relay (e.g., σ, as depicted in Fig. 10); Δ_i is the energy gap between the donor orbital and the relay orbital φ_i; $t_{i,i+1}$ is the strength of the interaction between the adjacent relay orbitals, φ_i and φ_{i+1}; Δ_1 is the energy gap between the donor orbital and the bridge orbital, φ_1, with which it interacts most strongly. Note that, as with equation (9), the splitting energy, $\Delta E(\pi)$, depends on the product of $n - 1$ intra-relay coupling terms, $t_{i,i+1}$.

It is crucial to note from equation (22) (and equation (9)) that $\Delta E(\pi)$ is a signed quantity. If the orbital basis set is chosen such that all adjacent orbitals overlap in-phase, as depicted in Fig. 10, then the matrix elements, $t_{i,i+1}$, $T_{D,1}$, and $T_{n,A}$, are all negative quantities. The signs of Δ_i are all positive. With these sign conventions, it follows from equation (22) that, for relays possessing an even number of bridge sites (i.e., orbitals), n, the splitting energy $\Delta E(\pi)$ is positive, whereas it is negative for relays having an odd number of bridge sites. For bridges possessing multiple relays, the net splitting energy is the sum of the splitting energies for the individual relays.[123] This summation leads naturally to the parity rule of TB coupling.[14,36,37] This rule states that the net coupling through two or more relays having the same parity of the number, n, of relay orbitals – that is, n is either even for all relays or odd for all relays – is given by the sum of the absolute magnitude of the splitting energy associated with each relay and is, therefore, strengthened by what may be called *constructed interference*. In contrast, the net coupling through relays having opposite parities is given by the difference in the absolute magnitude of the coupling associated with each relay and is, therefore, diminished by *destructive interference*.

This rule is illustrated for **15(4)** in Fig. 21. In this system the π MOs interact with two main relays (Fig. 21a and b), each possessing four C–C bonds, although overlap arguments suggest that the π-bridge coupling probably takes place with the allylic C–C bonds of each relay, rather than with the terminal bonds. This detail is, however, immaterial to the following analysis. Both main relays in **15(4)** have the same parity (even) and the net splitting energy is, therefore, double that for a single relay. This is, of course, the result predicted above for non-interacting relays. Because the two main relays are spatially close to each other, one must consider coupling pathways that cross from one main relay to the other, such as those depicted in Fig. 21c and d. In Fig. 21c, the coupling passes through a ring fusion bond connecting the two relays. This type of relay (there are two of them within the bridge depicted in Fig. 21) has an odd parity and the associated splitting energy is therefore negative. Consequently, the net negative splitting energy arising from these two relays counteracts the net positive splitting energy associated with the two main relays depicted in Fig. 21a and b, although the former is weaker than the latter,

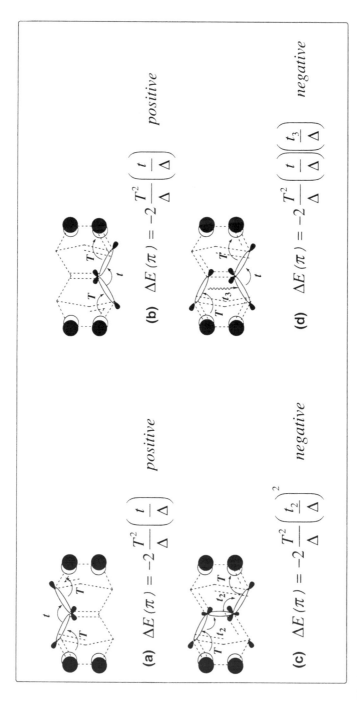

Fig. 21 Some through-bond coupling pathways in **15(4)**. The t_3 matrix element is responsible for the through-space interactions (represented by a wavy line). The McConnell splitting energy contribution from each pathway is given, as are the signs of the interactions. Note that t_i is negative for all values of i.

(a) $\Delta E\left(\pi\right) = -2\,\dfrac{T^2}{\Delta}\left(\dfrac{t}{\Delta}\right)$ *positive*

(b) $\Delta E\left(\pi\right) = -2\,\dfrac{T^2}{\Delta}\left(\dfrac{t}{\Delta}\right)$ *positive*

(c) $\Delta E\left(\pi\right) = -2\,\dfrac{T^2}{\Delta}\left(\dfrac{t_2}{\Delta}\right)^2$ *negative*

(d) $\Delta E\left(\pi\right) = -2\,\dfrac{T^2}{\Delta}\left(\dfrac{t}{\Delta}\right)\left(\dfrac{t_3}{\Delta}\right)$ *negative* on account of the odd-parity relays each being longer than the main relays by one bond. An alternative type of inter-relay coupling pathway incorporates a TS interaction, t_3, between two C–C bonds, one from each main relay, as shown in Fig. 21d. There are four pathways of this type. The splitting energy for each of these four coupling pathways is negative, because it involves an odd number of orbitals, or interactions; it is weak, however, since t_3 is numerically smaller than either t or t_2, but the total splitting energy arising from all pathways of this type is significant because these pathways increase in number with increasing bridge length. One could extend this analysis to coupling pathways incorporating multiple inter-relay jumps, but the qualitative conclusions remain unchanged.[41,122,123]

Our analysis leads us to conclude that inter-relay pathways in $15(n)$, $24(n)$, and $25(n)$ involve either an odd number of TS jumps or an odd number of ring fusion bonds, with the consequence that they interfere destructively with the coupling proceeding directly through the main relays, and this results in both the total splitting energy, $\Delta E(\pi)$, and β_{el} being degraded (that is, β_{el} increases in magnitude). This conclusion is consistent with the trends in the calculated magnitudes of $\Delta E(\pi)$ and β_{el} for $15(n)$, $24(n)$, and $25(n)$ (Fig. 20). For example, the finding that the magnitudes of β_{el} and $\Delta E(\pi)$, respectively decrease and increase along the series $24(n)$, $25(n)$, $15(n)$, is readily understood by noting that: (1) the number of destructive interference pathways of the type depicted in Fig. 21c decreases along the series $24(n) > 25(n) > 15(n)$; (2) the strength of the splitting energy associated with each of the destructive interference pathways of the type depicted in Fig. 21d also decreases along the same series, because the diminishing number of cyclobutane rings leads to a corresponding decrease in the numerical value of t_3 (i.e., the relays become progressively further apart along the series).

Can this simple and intuitive analysis of interference effects in TB coupling be advantageously used to design systems in which destructive interference is minimised or, better yet, in which the interference is predominantly *constructive*? The answer to this question is in the affirmative, and several such systems have been devised, two of which are $27(n)$ and $28(n)$ (Fig. 20). In contrast to the "standard" polynorbornane diene series, e.g., $25(n)$ (Fig. 22a), all reasonable inter-relay coupling pathways in $27(n)$ and $28(n)$ have the same parity as the main relays, namely even, in the case of the former (Fig. 22b) and odd, in the case of the latter (Fig. 22c). Consequently, on the basis of the parity rule, we predict that the net TB coupling in both $27(n)$ and $28(n)$ should be superior to that in $25(n)$, even though all three systems possess the same basic type of bridge. Gratifyingly, HF/3-21G calculations confirm this prediction. The calculated β_{el} values for $27(n)$ and $28(n)$ are *ten times smaller* than that for $25(n)$ and the coupling for a given bridge length is greater for the former pair of dienes, than for $25(n)$ (Fig. 20). For example, $\Delta E(\pi)$ for both $27(10)$ and $28(10)$ is ten times larger than that for $25(10)$ and this translates into a predicted two orders of magnitude rate enhancement for HT in the radical cations of $27(n)$ and $28(n)$, relative to $25(10)$.[41,123] To date, the prediction of reduced β_{el} values and amplified $\Delta E(\pi)$ splitting energies in $27(n)$ and $28(n)$, which was confirmed using higher levels of theory,[124] has yet to be verified experimentally.

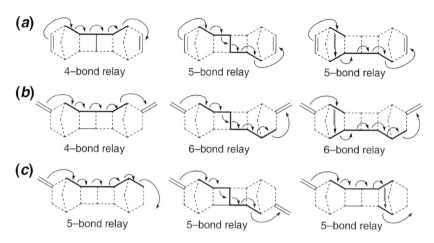

Fig. 22 Schematic showing some nearest neighbour coupling pathways for (a) **25(6)**, (b) **27(6)** and (c) **28(6)**. All significant pathways have the same parity in **27(n)** (even parity) and in **28(n)** (odd parity).

Generalisations concerning the design of bridges displaying enhanced coupling due to constructive interference effects are detailed elsewhere.[41,123]

THE DISSECTION OF TB COUPLING INTERACTIONS

As mentioned earlier, KT/3-21G calculations offer a simple computational method for reliably estimating both the distance dependence and the strength of the electronic coupling element for ET in a variety of systems.[39,83] In this context, one also requires a method for analysing and dissecting intra-bridge orbital interactions, and several elegant treatments are available.[39,42–45,84,85,89,125–140]

One such treatment, originally developed by Heilbronner and Schmelzer,[141] has proven to be most useful in analysing TB and TS coupling mechanisms. It relies on the adoption of a localised orbital representation and I shall give a brief description of it here. We,[39,84,85,89,129–131,133,137] Newton and co-workers,[7,45,126,127] and Miller and co-workers[125,138–140] have made use of the NBO scheme of Weinhold and coworkers.[88] NBOs are orthogonalised localised orbitals that conform closely to chemists' intuitive conceptions of localised orbitals (i.e., core, 2-center bonding and antibonding, lone pair, and Rydberg orbitals).

The NBO analysis of a molecule begins with the formation of the NBOs from the molecule's Hartree–Fock delocalised canonical MOs (CMOs), according to a specified localisation scheme. The Fock matrix, F^N, in the basis of the NBOs is constructed. This matrix, of course, is not diagonal, as is the matrix, F^C, in the basis of the CMOs. The off-diagonal matrix elements of F^N represent the strengths of the interactions between various NBOs and the diagonal elements correspond to the NBO self-energies. Diagonalisation of the full F^N matrix gives the CMOs and the CMO eigenvalues. The major advantage of the localised orbital approaches stems from the ability to construct *partial F^N* matrices which contain only a subset of the NBO interactions. Diagonalisation of such a partial matrix reveals how those NBO interactions which were retained in the matrix (as off-diagonal matrix elements) specifically influence the π, π and π^*, π^* orbital splitting energies by TB and TS coupling. In this way, a detailed picture of the relative contributions made by different relays of bridge NBOs may be developed.

Figure 23 illustrates the application of the NBO analysis method to a simple 4-orbital model, consisting of two π NBOs (π_1 and π_4) and the σ_2 and σ_3^* NBOs of the central C–C bond. The F^N matrix is first made blank, by switching off (i.e., zeroing) all off-diagonal elements, while retaining the diagonal ones, which are the self-energies of the π_1, π_4, σ_2, and σ_3^* NBOs (Fig. 23). The $\langle \pi_1 | F | \pi_4 \rangle$ off-diagonal matrix element is then switched on and diagonalisation of the resulting 2×2 submatrix gives the delocalised orbitals, $\pi_+ = (\pi_1 + \pi_4)$ and $\pi_- = (\pi_1 - \pi_4)$, and the resulting splitting energy provides a measure of the strength of the TS interaction between these two π NBOs. In the next step, mixing with σ_2 is included and this TB interaction causes the π_+ level to be raised above the π_-, level, leading to the so-called inverted sequence of orbitals.[141] (By symmetry, π_- cannot mix with σ_2.) Finally, switching on the TB interaction with σ_3^* results in a small stabilisation of π_-. This simple case study reveals how NBOs can be used to estimate the separate contributions from TS and TB interactions, and also how they can be used to estimate the relative importance of different TB coupling pathways; its application to larger systems is straightforward. It is, indeed, a powerful qualitative tool for probing orbital interactions!

6 ET mediated by polyunsaturated bridges

UNSATURATED HYDROCARBON BRIDGES

We now turn briefly to discuss the distance dependence of superexchange-mediated ET occurring through polyunsaturated hydrocarbon bridges. From the admittedly crude relationship between β_{el} and the donor–bridge energy gap, Δ, expressed by equation (10), we predict that increasing the degree of unsaturation in the hydrocarbon bridge will lead to smaller β_{el} values because Δ will diminish and the absolute magnitude of t will increase. The introduction of isolated double bonds into the bridge should cause a moderate decrease in the magnitude of β_{el}, whereas extended conjugation should obviously have a dramatic effect because the π and π^* manifolds of a conjugated bridge will lie close to the chromophore π and π^* levels.

This trend is nicely revealed by the ET distance dependence studies carried out on the SAMs of unsaturated bridges shown in Fig. 24. The β value for ET mediated by the saturated polymethylene bridge, determined from conducting probe atomic force microscopy, is 0.94 Å^{-1} (Fig. 24a).[121] (β for this system is also given in Fig. 19e, but in units of bond^{-1}; all β values listed in Fig. 24 are in Å^{-1} units because they were reported in this form in the literature.). A similar β value of 0.90 Å^{-1} was

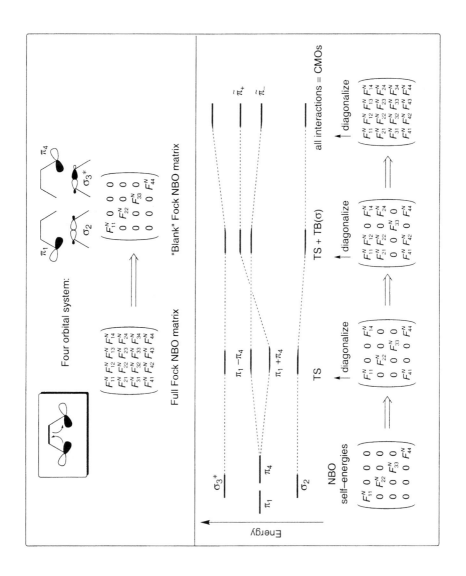

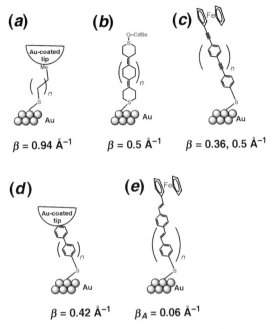

$\beta = 0.94$ Å^{-1} $\beta = 0.5$ Å^{-1} $\beta = 0.36, 0.5$ Å^{-1}

$\beta = 0.42$ Å^{-1} $\beta_A = 0.06$ Å^{-1}

Fig. 24 β values for the distance dependence of ET rates taking place in self-assembled monolayers (SAMs). (a) ET rates were determined using conducting probe atomic force microscopy.[121] (b) and (d) Photoinduced ET rates were determined using systems made from chemisorbing colloidal CdSe quantum dots onto gold-based SAMs.[144,145] (c) Two different determinations of the distance dependence of interfacial ET rates gave two different β values, 0.36 per Å[143] and 0.5 per Å.[146] (e) Interfacial thermal ET rates through oligo-p-phenylenevinylene bridges were obtained using the indirect laser-induced temperature jump method. Note that the β_A value refers to the distance dependence of the Arrhenius pre-exponential factor, A : $k_{et} = A \exp(-E_a/RT)$.

determined from interfacial ET rate measurements of a SAM comprising a polymethylene chain capped by a ferrocene group.[142,143]

A β value of 0.5 Å^{-1} was obtained for photoinduced ET taking place in a SAM between a gold electrode and a colloidal CdSe quantum dot, both attached to a bridge containing non-conjugated double bonds (Fig. 24b).[144,145] The double bonds in this system presumably lower the donor–bridge energy gap, and this leads to a

---

Fig. 23 An illustration of the use of the NBO (or any localised orbital) procedure for analysing TB and TS interactions, using as an example, butane-1,4-diyl. The model includes two chromophore π NBOs, π_1 and π_4, and the σ_2 and σ_3^* NBOs of the central C–C bridge bond. Firstly, the full Fock matrix, F^N, in the basis of the NBOs is constructed, and the off-diagonal matrix elements are then deleted, to form a blank Fock matrix (top part of the figure). In the bottom part of the figure, the Fock matrix is built up, starting with the blank matrix, and adding, in succession, the TS interaction between π_1 and π_4, the TB interaction with σ_2, and, finally, the TB interaction with σ_3^*, which produces the π CMOs.

smaller value for β, compared to that for a fully saturated bridge (Fig. 24a). Introduction of conjugation into the bridge further reduces the value of β (Fig. 24c and d). Two different β values have been reported for interfacial ET in SAMs containing the oligo-p-phenylene-ethynylene bridge (Fig. 24c), although the smaller value[143] would appear to fit the trend in the β values depicted in Fig. 24 better than the larger value.[146] That the β value of 0.42 $\mathring{A}^{-1}$ for the oligo-p-phenylene bridge (Fig. 24d) is not much smaller than that found for the non-conjugated unsaturated bridge (Fig. 24b) could be due to the small degree of conjugation in the former bridge (the adjacent phenyl groups are twisted out of coplanarity by about 40°).

In contrast, the oligo-p-phenylenevinylene bridge is completely planar and so conjugation is maximised in this entity. Interfacial thermal ET taking place through this bridge between a gold electrode and a covalently linked ferrocene group has been studied using the indirect laser-induced temperature jump method (Fig. 24e).[147] Arrhenius plots – from $k_{et} = A \exp(-E_a/RT)$ – gave activation energies, E_a, and pre-exponential factors, A, for ET in this system. The distance dependence of the pre-exponential factor was found to follow an extremely weak exponential decay, with a β_A value of only 0.06 $\mathring{A}^{-1}$ (Fig. 24e). Notwithstanding this very small β_A value, an electron transport mechanism involving conduction or hopping through the bridge (see Section 3) was ruled out on the grounds that the bridge virtual states are significantly higher in energy than the donor and acceptor states, by at least 0.8 eV.[147] Instead, the authors conclude that interfacial ET in their systems occurs adiabatically, rather than non-adiabatically, on the grounds that the superexchange coupling between the redox pair and the oligo-p-phenylenevinylene bridge is much larger than 200 cm^{-1} (e.g., see Fig. 5a). This is to be expected because, from the point of view of the superexchange mechanism, the 0.8 eV energy gap between the chromophore and bridge states is small and it should lead to strong superexchange coupling (e.g., see equation (9)). As a consequence of their adiabatic nature, the interfacial ET rates in the oligo-p-phenylenevinylene systems are controlled, not by the electronic coupling matrix element, V_{el}, but by structural reorganisation. The extended conjugation in the oligo-p-phenylenevinylene bridge facilitates rapid ET through the bridge by a particularly efficient superexchange mechanism, taking place in less than 20 ps for bridge lengths extending to 28 $\mathring{A}$.

Under the more energetic condition of photoinduced ET, molecular wire behaviour has actually been observed for the oligo-p-phenylenevinylene bridge. The systems, shown in Fig. 25, comprise a tetracene (TET) donor and a pyromellitimide (PI) acceptor covalently linked to oligo-p-phenylenevinylene bridges of lengths ranging from one aromatic ring, in **29(1)**, to five, in **29(5)**.[148] Rates of photoinduced charge separation, from locally excited TET to PI, were measured for this series in methyltetrahydrofuran. Two distinct ET behaviours were observed (Fig. 25, lower left-hand inset). For the two shortest members of the series, the charge separation rate decays with a β value of 0.27 $\mathring{A}^{-1}$. This value is typical of a superexchange mechanism involving a short conjugated bridge (cf. Figs. 24c and d).

In contrast, there is an abrupt change in the ET mechanism beginning with **29(3)**, for which the charge separation rate is greater than that for **29(1)**, even though the bridge in the former system is 13 Å longer than in the latter. Moreover, an extremely small β value of only 0.04 Å^{-1} is obtained for the longer members of the series **29(3)**–**29(5)**. The distance decay characteristics of the charge separation rates for **29(3)**–**29(5)** were interpreted in terms of molecular wire behaviour; in these systems, the energy of the bridge LUMO approaches that of the TET chromophore (see Fig. 25, lower right-hand inset) and thermal injection of an electron, from locally excited TET, into the bridge becomes energetically feasible for **29(n; n > 2)**. For the first two members of the series, this energy gap is too large for thermal electron injection into the bridge to take place; consequently, ET occurs by a superexchange mechanism in **29(1)** and **29(2)**.[148]

PROTEINS

Long-range ET in redox proteins is important and much work has been carried out to determine and understand the distance dependence of ET rates in both natural redox proteins and genetically modified ones, and in synthetic polypeptides. A substantial number of experimental studies, carried out over the past two decades, have produced β values that generally lie within the range 0.8–1.4 Å^{-1}.[1,72–75,149,150] Thus, the distance dependence behaviour of ET rates mediated by protein bridges is similar to that mediated by saturated hydrocarbon bridges. This similarity should not be too surprising because proteins are pretty much saturated entities possessing high lying virtual ionic states. The magnitude of the β values for proteins points to a superexchange mechanism operating in protein-mediated ET.

Analysing superexchange coupling in proteins is not as straightforward as it is for hydrocarbon bridges because there are, in general, hundreds, maybe even thousands, of different electron tunnelling pathways, of comparable strengths, between the redox centres in a protein; consequently, they should all be considered in any theoretical treatment of electronic coupling. In addition, these pathways contain a diverse range of bonded and non-bonded (i.e., H-bonded) intra-relay interactions that renders application of the simple McConnell equation (9) inadmissible. A generalised McConnell-based model of tunnelling pathways in proteins has been developed which is quite successful in analysing, and assessing the relative importance of, tunnelling pathways.[151–154] This tunnelling pathway model represents the different types of interactions between atoms within the protein in terms of three types of electronic coupling matrix elements, namely covalent interactions, t_C, interactions through H-bonds, t_{HB} and TS interactions, t_{TS}. The values of these interactions are determined empirically. The contribution to the overall electronic coupling by the mth pathway, P_m, is given by the products of the various interactions, t, between the components of that pathway, i.e.,

$$V_{el}(P_m) = \prod_i t_C(i) \prod_j t_{HB}(j) \prod_k t_{TS}(k) \tag{23}$$

This type of analysis enables one to find the optimal tunnelling pathways within a particular protein.[155]

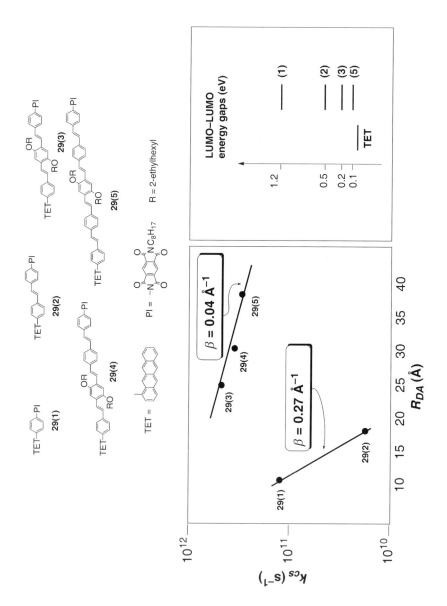

DNA

Electron transfer within the DNA duplex has been extensively reviewed recently and only a brief discussion will be given here.[28,30,31,156,157]

Two mechanisms have, to date, been experimentally identified for charge migration taking place through duplex DNA, namely superexchange-mediated ET[31] and charge hopping[30] – see Section 3. Whether ET and HT in DNA occur by charge hopping or by superexchange depends largely on the magnitude of the energy gap between the bridge ionic states (i.e., the DNA bases) and the donor state. If this gap is large, then the ET or HT process will take place by a superexchange mechanism, i.e., by a single coherent "jump", from the donor to the acceptor. The CT dynamics will then follow an exponential decay with increasing donor–acceptor distance, according to equation (7). If, on the other hand, the energy gap is of the order of $k_B T$, then thermal injection of an electron or a hole into the DNA bridge may take place, and charge transport will occur by a hopping mechanism, as described in Section 3. The hopping dynamics will display weak distance dependence, given by equation (8).[26,27,158] Thus, because of its exponential distance dependence, superexchange-mediated ET and HT in DNA should not be significant for donor–acceptor separations exceeding 15 Å, whereas ET and HT occurring by a hopping mechanism are expected to be much longer range processes, extending beyond 50 Å.

The composition and sequence of the base pairs in DNA should have an important effect on the ET and HT mechanisms. With regard to HT, guanine (**G**) is much more easily oxidised than either adenine (**A**) or thymine (**T**); consequently, hole injection into the DNA helix by a suitable donor will generate a $G^{+\cdot}$ radical cation. The hole may now randomly move over large distances along the DNA helix by hopping between adjacent **G** bases. The hole is less likely to reside on either **A** or **T** bases because of their unfavourable oxidation potentials (see below, however). The **A**–**T** base pairs serve, instead, as a superexchange medium through which the charge tunnels between adjacent **G** bases. The rate of hopping between a pair of adjacent **G** bases therefore follows an exponential dependence on the number of intervening **A**–**T** base pairs. The overall HT hopping rate between donor and acceptor is thus determined by the longest sequence of **A**–**T** base pairs between any pair of adjacent **G**bases in the charge transport pathway. If each hopping step in an HT pathway involves the same number of intervening **A**–**T** base pairs, then the hopping mechanism obeys the distance dependence relationship given by equation (8).[159]

Fig. 25 The series of dyads, **29**(*n*), possessing the oligo-*p*-phenylenevinylene bridge that were used to investigate the switchover from superexchange to molecular wire behaviour in the photoinduced electron transfer reaction, from the locally excited state of tetracene (TET) donor to the pyromellitimide (PI) acceptor.[148] Also, shown are a schematic of the photoinduced charge separation rate versus, donor–acceptor distance (lower left-hand side) and the LUMO energies of TET and the various bridges (lower right-hand side).

In an elegant series of experiments, Giese and co-workers have investigated the hole hopping mechanism between **G** bases in duplex DNA.[27,28,30,158–161] In these experiments, Norrish I photocleavage of an acylated nucleoside in a synthetic duplex DNA leads to the formation of a sugar radical (Fig. 26), which then undergoes heterolysis to give the radical cation. This sugar radical cation initiates the HT process by transferring the positive charge to the nearby **G** base (i.e., G_{23} in Fig. 26). The hole is then free to leave the initial site, $G_{23}^{+\cdot}$, and wander through the DNA molecule, hopping between adjacent **G** bases by a superexchange mechanism, as depicted in Fig. 27. This hole hopping is only terminated when the hole encounters a **GGG** triplet and is irreversibly trapped (**GGG** has a lower oxidation potential than **G**). In the presence of water, both the $G_{23}^{+\cdot}$ radical cation and the **GGG**$^{+\cdot}$ triplet radical cation are trapped irreversibly to give readily identifiable cleavage products. The relative rate of hole transfer, $k_{ht}(\text{rel})$, from the $G_{23}^{+\cdot}$ radical cation to the **GGG** triplet, is given by the ratio of the cleavage products at the **GGG** triplet and the single **G** base.[28] By varying the length, composition and sequence of the π-stack spanning $G_{23}^{+\cdot}$ and **GGG**, the distance dependence of the HT dynamics was obtained (Fig. 27).

For **30(1)**–**30(4)** (Fig. 27a), in which one to four **A**–**T** base pairs are inserted between $G_{23}^{+\cdot}$ and **GGG**, the relative HT rate, from $G_{23}^{+\cdot}$ to **GGG**, was found to follow an exponential decay with increasing G_{23}–**GGG** separation, with a β value of $0.7\ \text{Å}^{-1}$. The superexchange mechanism, mediated by the intervening **A**–**T** base pairs is clearly operating here, and the smaller magnitude of β, compared to that for saturated hydrocarbon bridges and proteins, reflects the unsaturated character of the **A** and **T** bases (i.e., their virtual cationic states lie closer in energy to the $G_{23}^{+\cdot}$ level than do those of saturated hydrocarbon bridges). Replacement of one of the four **A**–**T** base pairs in **30(4)** with a **G**–**C** base pair, to give **31(4)** and **32(4)**, led to a two orders of magnitude increase in the HT rate relative to **30(4)**. This rate enhancement was attributed to reversible hole hopping between $G_{23}^{+\cdot}$ and the inserted **G** unit.[158] The longest superexchange pathway in both **31(4)** and **32(4)** has been reduced to only two **A**–**T** base pairs, compared to four in **30(4)**. It is not surprising, therefore, that the HT rates for **31(4)** and **32(4)** are larger than that for **30(4)**, and are

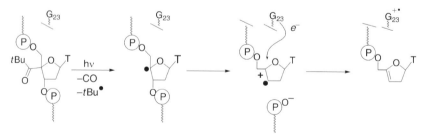

Fig. 26 Hole transfer propagation through DNA.[28,158,160] The hole is generated by Norrish I photocleavage of an acylated nucleoside, followed by heterolysis, to give the sugar radical cation. The hole is then injected into a nearby **G** base.

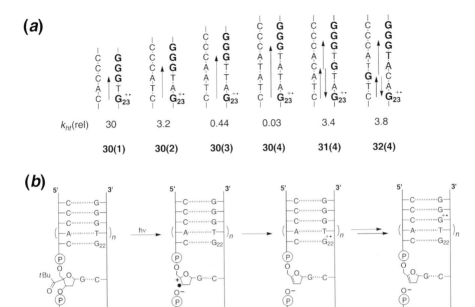

Fig. 27 (a) Relative HT rates, k_{ht} (rel), from $G_{23}^{+ \cdot}$ to **GGG**, through various DNA duplexes, **30(n)**–**32(n)**, where n is the number of base pairs between G_{22} and **GGG**. The shaded areas highlight the longest path between adjacent **G** bases. (b) Hole injection into the single G_{22} base, followed by hole transport to the **GGG** triplet in a DNA duplex possessing a variable number, n, of **A**–**T** base pairs; $n = 1$–5, 7, 8, 16.

comparable to that for **30(2)**, which also has a superexchange pathway extending over two **A**–**T** base pairs.

By using synthetic duplex DNA molecules in which every hopping step between adjacent **G**–**C** pairs involves the same number (two) of intervening **A**–**T** pairs, the validity of equation (8), with $\eta = 1.7$ was demonstrated.[160]

The distance dependence of the relative rates of HT in a series of DNA duplexes, **35(n; n = 1–5, 7, 8, 16)** (Fig. 27b), possessing a variable number, n, of **A**–**T** base pairs between the G_{22} base and the **GGG** triplet has recently been reported.[29] The distance between G_{22} and **GGG** in **35(n)** ranges from 7 Å ($n = 1$) to 60 Å ($n = 16$). Intriguingly, the HT rate in **35(n)** was found to display two distinct distance dependence behaviours. For short G_{22}–**GGG** separations ($n = 1$–3), the HT rate follows an exponential decay with increasing separation, with a β value of 0.6 Å^{-1}; thereby demonstrating that the superexchange mechanism is operating over this range of distances. In contrast, the HT rates become practically distance independent over larger G_{22}–**GGG** separations ($n > 4$). Clearly, the coherent superexchange mechanism is no longer operating over these distances; instead, a thermally induced hole hopping process takes place over long **A**–**T** sequences, in which the adenine bases are acting as hole carriers (**A**-hopping). The observed mechanistic switchover from superexchange to **A**-hopping occurs because the superexchange-mediated HT rate decreases rapidly with increasing $\mathbf{G_{22}}$–$\mathbf{GGG}$ distance and it soon reaches a point ($n \approx 4$) where it becomes slower than the *endothermic*[162] hole hop from $\mathbf{G_{22}^{+}}$ to the nearest **A** base. (From the oxidation potentials of **G** and **A**, 1.24 and 1.69 V, respectively,[31] the $\mathbf{G_{22}^{+}}$ to **A** hole hop is endothermic by about 43.5 kJ/mol.) Once injected into the first **A**–**T** base pair, the hole rapidly migrates (hops) along the **A**–**T** sequence. This type of mechanistic switchover in the distance dependence of HT rates through DNA duplexes was predicted from theoretical models.[18,20,163–165]

Electron transport through DNA, in which electron hopping involves radical anions of base pairs, is less likely to occur than hole hopping, involving radical cations, because the DNA bases have weak electron affinities;[166] consequently, the energies of the anionic states of base pairs, in general, lie well above those of conventional donor and acceptor chromophores.

A large number of distance dependence studies on photoinduced ET and hole transfer processes through DNA has been carried out using synthetic DNA duplexes containing redox chromophores that are attached to the duplex, either by non-covalent bonds (i.e., by intercalation), or by short covalently linked tethers. In these investigations, the data were found to be consistent with the superexchange mechanism since the ET and HT rates followed exponential decays with increasing distance.

An example of this type of distance dependence experiment is shown in Fig. 28.[31,167,168] The DNA strand is either a synthetic 6-mer or 7-mer duplex comprising **A**–**T** base pairs and a single **G**–**C** base pair located at various positions along the helix. The duplex is capped by a stilbene hairpin which serves as a photo-oxidant (Fig. 28a). Photoexcitation of the stilbene group initiates ET from the **G** donor to the locally excited stilbene acceptor (or equivalently, HT occurs from the locally stilbene to **G**). Importantly, neither **A** nor **T** is able to transfer an electron to the locally excited stilbene since their oxidation potentials are too high. Two different series of DNA hairpins were studied; in one series, **nG:C**, the **G** base is attached to the **T**-bearing strand (Fig. 28b), and in the other series, **nC:G**, the **G** base is attached to the **A**-bearing strand (Fig. 28c).

By varying the stilbene–**G** separation it was possible to obtain a distance dependence for both the photoinduced rate of charge separation (CS) and the subsequent rate of charge recombination (CR) for each series of DNA hairpins.[167]

Fig. 28 Photoinduced electron transfer studies carried out on 6-mer (not shown) and 7-mer hairpin DNA duplexes capped by a stilbene acceptor chromophore.[31,167,168] The duplexes contain **A**–**T** base pairs and a single **G**–**C** base pair, whose position in the duplex is varied. (a) The photoinduced ET process is illustrated for **3G:C**, in which the **G**–**C** base pair is third removed from the stilbene (St) group. The stilbene fluorescence is quenched by electron transfer from the **G** donor. (b) Damping factors for charge separation, β(CS), and subsequent charge recombination, β(CR), for *n* **G:C**, in which the **G** base is connected to the **T**-bearing strand. (c) Damping factors, β(CS) and β(CR), for *n* **C:G**, in which the **G** base is connected to the **A**-bearing strand.

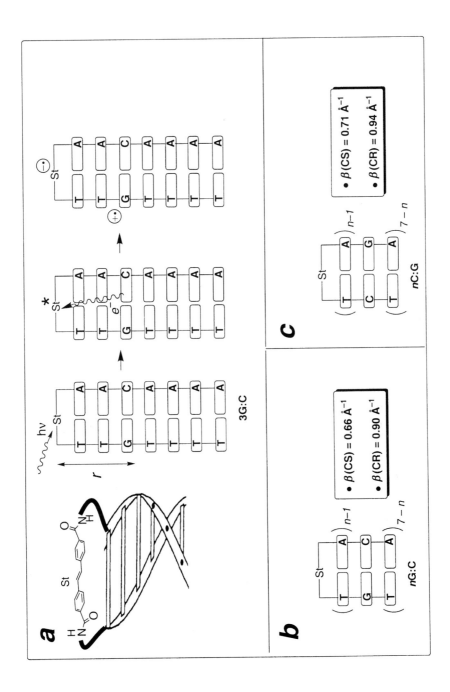

The rates all followed good exponential decay profiles with increasing stilbene$- >$ G separation, with the following β values:

nG:C, β (CS) $= 0.66$ Å^{-1} and β (CR) $= 0.90$ Å^{-1} (Fig. 28b)
nC:G, β (CS) $= 0.71$ Å^{-1} and β (CR) $= 0.94$ Å^{-1} (Fig. 28c)

The β values for charge separation are significantly smaller than those found for saturated hydrocarbon bridges, whereas those for charge recombination are comparable to the latter. The substantial difference in the magnitudes of the β values for CS and CR is readily explained by the inverse dependency of β on the energy gap, Δ, separating the chromophore and bridge states – see equation (10). For the charge separation process, Δ is estimated to be 0.2 eV, whereas for charge recombination process, it is much larger, about 0.5 eV.[167]

Various distance dependence studies of superexchange-mediated ET and HT rates through duplex DNA, involving different base-pair sequences and different redox couples, have produced β values ranging from 0.60 to 1.0 Å^{-1}. Thermal HT processes gave β values of 0.6 and 0.7 Å^{-1}(see earlier)[29,158] and the hairpin experiments gave β values of 0.68 Å^{-1}(CS) and 0.92 Å^{-1}(CR).[167] Kelley and Barton[169] report a β of 1.0 Å^{-1}, based on fluorescence quenching studies. Harriman and co-workers[170,171] have studied the distance dependence of photoinduced ET rates between non-covalently attached intercalated donor and acceptor groups in duplex DNA and they obtained β values of about 0.9–1.0 Å^{-1}. A similar estimate of $\beta \approx 1.0$ Å^{-1} was obtained by Olson and co-workers.[172]

Interestingly, Fukui and co-workers reported an anomalously larger β value of 1.5 Å^{-1}, determined from the distance dependence of photoinduced ET, taking place from a **G−C** base pair to a tethered, intercalated, locally excited acridine acceptor group.[173,174] This system has been recently re-investigated using femtosecond pump-probe spectroscopy.[175–177] It was found that, with a **G−C** base pair as the nearest neighbour to the acridine acceptor, the photoinduced ET process took place extremely rapidly, within 4 ps. The placement of one **A−T** base pair between the acridine and **G−C** resulted in a dramatic decrease in the ET rate, giving an apparent β value of >2.0 Å^{-1}. This huge decrease in the ET rate was attributed to a change in the ET mechanism, from one that is nearly activationless, to one that requires thermal activation. That is, the distance dependence of the rate is due, not only to a decrease in the strength of the electronic coupling, V_{el}, with increasing donor−acceptor distance, but also to a concomitant increase in the activation energy. The Fukui−Tanaka intercalated acridine−DNA system is a classic example of the important effect that Franck−Condon factors may exert on the magnitude of the phenomenological β value.[177]

7 A summary of β values

It is informative to list and compare the β values for the distance dependence of superexchange-mediated ET rates through the various types of bridges reviewed in this article. For consistency, the units of β are given in Å^{-1}.

Saturated hydrocarbon bridges, $\beta \approx 0.75-0.98$ Å^{-1}.   (24)

Unsaturated, unconjugated hydrocarbon bridges, $\beta \approx 0.5$ Å^{-1}.   (25)

Conjugated hydrocarbon bridges, $\beta \approx 0.04-0.40$ Å^{-1}.   (26)

Proteins, $\beta \approx 0.8-1.4$ Å^{-1}.   (27)

Duplex DNA, $\beta \approx 0.60 - 1.0$ Å^{-1}.   (28)

The fairly wide dispersion of the β values for DNA reflects the sensitivity of β to the energy gap, Δ, between the chromophore and bridge states, as expressed by equation (10). The magnitude of Δ is expected to be quite small (~ 0.5 eV) because the π and π^* ionic states of the DNA base pairs are close in energy to the chromophore states. Hence, β is sensitive to any small variation in the magnitude of Δ, brought about primarily by variations in the energies of the various redox chromophores that have been used. The broad range of β values found for proteins probably has a different origin: in these complex structures the superexchange pathways comprise mixtures of both bonded and non-bonded contacts. Because these two types of contacts are associated with different values of t interactions (e.g., equation (23)), the magnitude of β will depend on the relative proportions of these contacts within the superexchange pathways. Since the relative proportions of these contacts vary from one redox protein to the next, the observed broad range of β values is to be expected.

From the data given in equations (24)–(28), we may conclude that super-exchange-mediated ET through the DNA duplex is significant and that β for DNA may be comparable to the upper limit of β observed for conjugated hydrocarbon bridges. However, it now seems clear that DNA does not behave as a molecular wire in ET reactions, as has been posited.[178]

Finally, the remarkably small β values – compared to that for TS ET – for ET through saturated hydrocarbon bridges poses an interesting question: What does one use as an electrical insulator in charge-bearing nano-devices that require electrical insulation? Fifteen years ago, the answer would have been to use saturated hydrocarbons since it is well known that they are excellent electrical insulators. Although they are, indeed, electrical insulators, even at the molecular level, they can, nevertheless, facilitate, by the superexchange mechanism, very rapid ET over distances exceeding 12 Å. Thus, they might well short-circuit charged nanodevices that have hydrocarbon molecules interspersed between them. One would need to separate the devices by at least 20 Å to prevent this short-circuit from occurring. Thus, there may be a lower limit to the separation between charged devices, below which they will exchange charges. This limit might be reduced by using hydrocarbons possessing a large number of *gauche* conformations which will reduce the electronic coupling by dint of the all-*trans* rule. The deployment of fluorcarbons might also serve to reduce the separation limit, because, on energetic grounds, the fluorine substituents render the C–C σ and σ^* manifolds less able to participate in a superexchange mechanism.

8 The singlet–triplet energy gap in CS states

CHARGE RECOMBINATION VIA THE TRIPLET MANIFOLD

I now return to our old friends, the series of dyads **18(n)** which has provided so much fundamental mechanistic information concerning CT processes. Figure 29 shows an energy diagram for the photophysical processes that we have observed for **18(n)**. Photoexcitation of the **DMN** chromophore leads to the formation of the singlet CS state, $^1\mathbf{D}^+\mathbf{A}-$, (Fig. 29), whose energy depends on the solvent polarity, but it always lies above that of the locally excited **DMN** triplet state, $^3\mathbf{D}^*\mathbf{A}$. Consequently, the $^1\mathbf{D}^+\mathbf{A}^-$ state may undergo charge recombination (CR) by two competing pathways, namely direct CR to the singlet ground state (k_{cr}^1), or by intersystem crossing (isc) to the triplet CS state, $^3\mathbf{D}^+\mathbf{A}^-$ (ω_{isc}), followed by CR to $^3\mathbf{D}^*\mathbf{A}$, and thence a

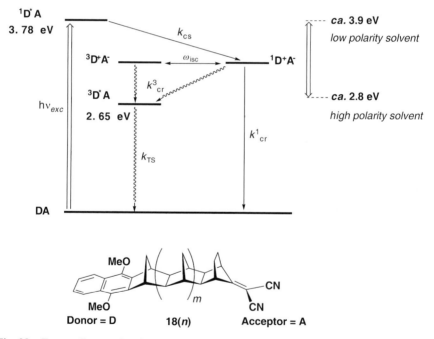

Fig. 29 Energy diagram for charge separation and charge recombination processes in **18(n)**. The energy of the CS state varies as a function of the bridge length, n, and the solvent polarity, from slightly above $^1\mathbf{D}^*\mathbf{A}$ (i.e., for $n = 12$ in saturated hydrocarbon solvents), to about 1 eV below (i.e., in highly polar solvents). Under all circumstances the locally excited donor triplet ($^3\mathbf{D}^*\mathbf{A}$) lies below the CS state.

spin-forbidden transition to the singlet ground state. Population of the locally excited triplet, $^3D^*A$, might also occur via direct isc, from the singlet CS state. The overall triplet recombination CR rate by these two isc pathways is represented by k_{cr}^3.

The rates of both singlet and triplet CR processes, k_{cr}^1 and k_{cr}^3, respectively, have been found to diminish exponentially with increasing bridge length, but with significantly different damping factors, β. For example, in 1,4-dioxane solvent, the damping factor,[3] β, for the triplet CR process is 0.56 bond^{-1}, whereas that for the singlet CR process, $^1\beta = 1.22$ bond^{-1}, is twice as large.[179]

The reason for the very different values of these damping factors may be explained by noting that $^3\beta$ for triplet CR is determined by two quantities, namely the product $|V_{el}|^2 \cdot$FCWD for the triplet CR rate (see equation (3)), and the frequency, ω_{isc}, of isc which determines the population of the triplet CS state.

For short bridge lengths, the isc frequency, ω_{isc}, is probably quite small because the scalar spin-exchange coupling interaction, J – which is one-half of the energy gap, ΔE_{ST}, between the singlet and triplet CS states – is expected to be quite large, in comparison with the spin–orbit coupling (SOC) and the electron-nuclear hyperfine interaction terms. The SOC and hyperfine terms are the perturbations which drive the isc process and they become more effective as they approach J in magnitude. For example, we have calculated that ΔE_{ST} is 53 cm^{-1} for the CS state of **18(6)**, compared to ~ 0.01 cm^{-1} for both SOC and hyperfine terms (see later).[180] Both ΔE_{ST} and the SOC interaction decay exponentially with increasing bridge length, whereas the hyperfine interaction remains unchanged. Consequently, ω_{isc} increases with increasing bridge length – due to the plummeting magnitude of ΔE_{ST}, which is only 0.44 cm^{-1} for the CS state of **18(12)**[181] – until it becomes larger than the singlet CR rate, k_{cr}^1. This increasing magnitude of ω_{isc} with increasing bridge length has an ameliorating effect on the magnitude of $^3\beta$ because it increases the triplet character of the CS state, thereby enhancing triplet CR. Furthermore, the Franck–Condon term, FCWD, for the triplet CR process should display a weaker distance dependence than that for the singlet CR process, on the grounds that the former process takes place in the Marcus normal region, whereas the latter occurs in the inverted region.[179]

In summary, the smaller $^3\beta$ value, compared to $^1\beta$, together with the fact that k_{cr}^3 exceeds k_{cr}^1 at longer bridge lengths, is due to the increased triplet character of the CS state at larger distances and the more favourable FCWD factor for the triplet CR pathway. Within this context, it should be pointed out that the β value of 1.1 bond^{-1}, given in equation (16), for thermal charge recombination in the CS states of **18(n)**, is a composite of both $^1\beta$ and $^3\beta$ because the TRMC method, which was used to obtain the lifetime data, measures the combined lifetimes of the triplet and singlet CS states.

THE DISTANCE DEPENDENCE OF ΔE_{ST}

The singlet&triplet energy gap, ΔE_{ST}, in organic biradicals[35,182,183] and (inter-molecular) radical ion pairs[184–190] has been shown to be a useful indicator of electronic coupling in ET and excitation energy transfer (EET) processes.[191–194]

This nexus arises from the fact that ΔE_{ST} for a biradical is equal to twice the scalar spin-exchange coupling, J, ($\Delta E_{ST} = 2J$) and this is discussed in more detail in Section 8.3. Although J is not identical to the electronic coupling matrix elements, V_{el}, for ET and EET, all three quantities should follow similar trends upon variation of parameters, such as bridge length, bridge configuration, and the orientation of the donor and acceptor groups. Indeed, J values have been determined from EPR spectra of a wide range of flexible polymethylene α, ω-acyl-alkyl and α, ω-bisalkyl biradicals.[195–198] These J values, while insightful, are, in fact, average values, $\langle J \rangle$, taken over the conformational space sampled by the flexible chain connecting the radical centers.[192,196,197] While investigations of flexible-chain biradicals have provided information concerning the relative importance of TS, through-solvent and through-bond coupling contributions to the magnitude of $\langle J \rangle$, it is desirable to explore such effects using rigid biradical species, in which separation and orientation between the radical centres are structurally well-defined and controllable.

This challenge has been met in the recent work of Wegner and co-workers,[181,199] who used the field-dependent photo-CIDNP method to determine the value of J for the CS states of two of our dyads, namely $^+$DMN[10]DCV$^-$ and $^+$DMN[12]DCV$^-$, generated, respectively, from solutions of **18(10)** and **18(12)** in benzene, or 1,4-dioxane, by intra-molecular photoinduced ET. The experimental J values for $^+$DMN[10]DCV$^-$ and $^+$DMN[12]DCV$^-$ (in benzene) are 1.06 and 0.22 cm^{-1}, respectively. These values are surprisingly large, in light of the large inter-chromophore separations in these systems (Scheme 8), and suggest that they are caused by a through-bond (superexchange) mechanism. Moreover, J was found to be positive for both systems, meaning that the triplet CS state is more stable than the singlet CS state – by 2.12 cm^{-1}, in the case of $^+$DMN[10]DCV$^-$, and by 0.44 cm^{-1}, in the case of $^+$DMN[12]DCV$^-$ (i.e., $\Delta E_{ST} = 2J$) (Fig. 30a).

With only two experimental points, it is impossible to determine the functional form of the distance dependence of ΔE_{ST}. Assuming an exponential distance dependence for ΔE_{ST} – in accordance with the superexchange mechanism – we obtain a β_{ST} value of 0.79 bond^{-1} for the singlet–triplet energy gap. We used time-dependent density functional theory (TD-DFT) to calculate the distance dependence of ΔE_{ST} for the entire series of CS states, $^+$DMN[n]DCV$^-$.[180] The TD-DFT ΔE_{ST} values do, indeed, follow an exponential decay with increasing bridge length (Fig. 30b), with a β_{ST} value of 0.92 bond^{-1}, which is fairly close to the experimental value of 0.79 bond^{-1}. The magnitude of β_{ST} confirms that superexchange is the principal cause of the singlet–triplet energy gaps in $^+$DMN[n]DCV$^-$.

Further evidence in support of the superexchange mechanism came from the finding that the TD-DFT calculated magnitude of ΔE_{ST} is sensitive to the configuration of the norbornylogous bridge, rapidly diminishing with increasing number of *gauche* conformations within the bridge, in accordance with the all-*trans* rule of TB coupling (see Section 4).[123] For example, the TD-DFT ΔE_{ST} value of 1.6 cm^{-1} for the CS state of **19(8)** (Fig. 16) is six times smaller than that for the CS state of **18(8)** (9.6 cm^{-1}).[180]

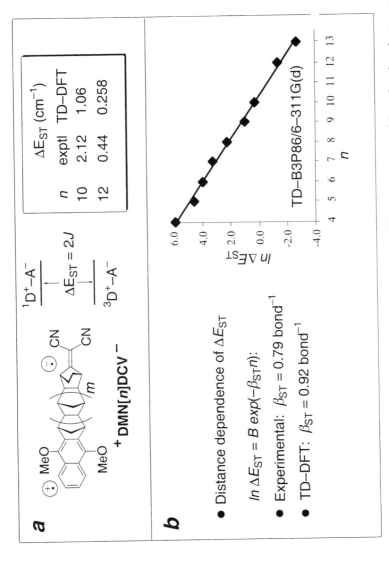

Fig. 30 (a) Experimental and calculated (TD-B3P86/6-311G(d)) singlet–triplet energy gaps, ΔE_{ST}, for the charge separated states, $^{+}$**DMN[n]DCV**$^{-}$. The experimental ΔE_{ST} values were determined in benzene solution. (b) The distance dependence of ΔE_{ST}. The graph is the distance dependence of ΔE_{ST} calculated at the TD-B3P86/6-311G(d) level of theory.

CONNECTIONS BETWEEN ΔE_{ST} AND ET PROCESSES

The experimental and calculated β_{ST} values of 0.79 and 0.92 bond^{-1} for the distance dependence of ΔE_{ST} for the $^+$**DMN[n]DCV**$^-$ series of CS states corresponds closely to the average experimental β value of 0.92 bond^{-1} for a variety of ET processes occurring in the same series of molecules, **DMN[n]DCV** = **18(n)**. Thus, the average values are:

$$\beta_{ST} = 0.85 \text{ bond}^{-1} \text{ and } \beta_{ET} = 0.92 \text{ bond}^{-1} \tag{29}$$

where the notation β_{et} is used in place of β, in order to stress that it refers to ET processes.

The similarity of the β_{ST} and β_{et} values is not coincidental, and it follows from the fact that both the non-adiabatic ET rate and ΔE_{ST} depend on the *square* of the electronic coupling element for ET, V_{el}. The $|V_{el}|^2$ dependence of the non-adiabatic ET rate follows from the Fermi Golden Rule, equation (3). The dependence of ΔE_{ST} on $|V_{el}|^2$ may be seen with the aid of Fig. 31, which shows the four relevant potential energy curves representing the locally excited singlet and triplet diabatic configurations, $^1(^*$**DMN[n]DCV**) and $^3(^*$**DMN[n]DCV**), and the singlet and triplet CS diabatic configurations, $^1(^+$**DMN[n]DCV**$^-$) and $^3(^+$**DMN[n]DCV**$^-$), which are assumed to be isoenergetic on the scale of the drawing and are represented by a single curve, labelled $^{1,3}(^+$**DMN[n]DCV**$^-$). The electronic coupling matrix elements $^1V_{cs}$ and $^3V_{cr}$ are associated, respectively, with the formation of the singlet charge separated state, and with charge recombination to the locally excited

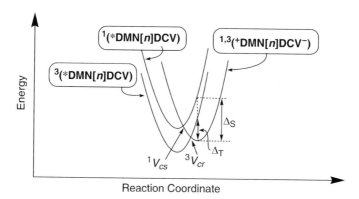

Reaction Coordinate

Fig. 31 Schematic of idealised parabolic diabatic potential energy curves versus a generic one-dimensional reaction coordinate which represents changes in both nuclear and solvent configurations. Two curves represent the singlet and triplet diabatic configurations, $^1(^*$**DMN[n]DCV**) and $^3(^*$**DMN[n]DCV**). The singlet and triplet CS diabatic configurations are assumed to be isoenergetic on the scale of the drawing and are represented by a single curve, labelled $^{1,3}(^+$**DMN[n]DCV**$^-$). The parabolas are placed such that both the singlet charge separation and the triplet charge recombination processes take place in the Marcus normal region, as is found experimentally.[92,179]

triplet state. The parabolas are placed such that both the singlet charge separation and the triplet charge recombination processes in **DMN[n]DCV** take place in the Marcus normal region, as is found experimentally.[92,179] The singlet–triplet splitting energy, ΔE_{ST}, *at the relaxed geometry of the CS states*, is given, to second order, by equation (30):[184,187,189]:

$$\Delta E_{ST} = \frac{|^3V_{cr}|^2}{\Delta_T} - \frac{|^1V_{cs}|^2}{\Delta_S} \tag{30}$$

where the energy gaps, Δ_T and Δ_S are defined at the relaxed geometries of the CS states, as depicted in Fig. 31. Making the assumption that $^3V_{cr}$ and $^1V_{cs}$ have identical distance dependencies, then the observed relationship $\beta_{ST} \approx \beta_{et}$ follows.

9 Spin-control of CS state lifetimes

Photoinduced charge separation is an important process because it leads to the transduction of light energy into useful chemical potential, as measured by the free energy change, ΔG_{cr}, for charge recombination (Fig. 32). These photo-generated CS states may therefore be regarded as molecular photovoltaic devices. A classic example of the usefulness of photoinduced charge separation in biology is the vital role it plays in the primary events of photosynthesis. The successful design of molecular photovoltaic devices rests on meeting three fundamental requirements, namely:

(1)　The quantum yield for the charge separation process should be as high as possible. That is, $k_{cs} \gg k_d$ (Fig. 32).
(2)　The mean lifetime of the CS state, τ_{cr} $(= 1/k_{cr})$, must be sufficiently long (μs–ms domain) for it to carry out "useful" chemical work.
(3)　The energy content of the CS state should be as high as possible, thereby maximising the conversion of photonic energy into chemical potential. Thus, $|\Delta G_{cs}|$ should be as small as practicable while ensuring that requirement (1) is met. This statement may be recast by saying that the energy conversion yield,

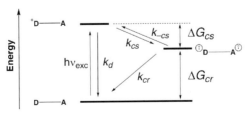

Fig. 32　Typical energy level diagram illustrating the photophysical processes that take place upon local excitation of the donor chromophore (a similar scheme obtains for excitation of the acceptor chromophore). k_d represents both radiative (fluorescence) and non-radiative decay processes.

defined as the ratio of the driving force for charge recombination $(-\Delta G_{cr})$ to the energy of the excitation photon, $h\nu_{exc}$, should be as close as possible to unity. Unfortunately, as ΔG_{cs} approaches zero, not only does the quantum yield for charge separation diminish, but back ET, from the CS state to the locally excited state – depicted by k_{-cs} in Fig. 32 – provides an additional decay route for the CS state, thereby shortening its lifetime.[200] For $-\Delta G_{cr} > 0.2$ eV, back ET is probably unimportant. In practice, the excitation energy, $h\nu_{exc}$, is generally greater than 2 eV for most systems; consequently, the optimal value for the energy conversion yield is about 0.9, i.e., $(-\Delta G_{cr})/h\nu_{exc} \approx 0.9$.

Considerable effort is being expended into prolonging the lifetimes of CS states. Clearly, this requires that the electronic coupling matrix element V_{cr} for CR is as small as possible. This may be achieved using a covalently linked dyad possessing a very long bridge. Indeed, we have demonstrated that the lifetimes of the photo-generated CS states of **18(n)** increase with increasing bridge length, and even reach 1 μs for the singlet CS state of **18(13)** (Fig. 15).[101,104] Unfortunately, increasing the bridge length in a dyad also leads to a decrease in the magnitude of the electronic coupling for photoinduced charge separation, and this results in a diminished quantum yield for photoinduced charge separation. For example, the quantum yield for formation of the CS state of **18(13)** is only *ca.* 30%. Thus, efficient charge separation and longevity of the resulting CS state have conflicting requirements in terms of optimal inter-chromophore separation, the former requiring short bridge lengths, and the latter long bridge lengths. This conflict may be referred to as the distance problem of charge recombination.

Two different approaches are currently being taken to circumvent this distance problem. The first, and more intensely studied approach is the construction of multichromophoric systems, i.e., triads, tetrads, pentads, etc., that constitute a gradient of redox centres arranged within a spatially well-defined array.[46,201–207] The principle behind this strategy is illustrated in Fig. 33 for a tetrad. In this type of system, the ET process takes place in a sequence of rapid "hops", between adjacent chromophores that are spanned by a bridge that is short enough to guarantee that each hop occurs with near unit efficiency. The final result is charge separation over a very large distance, often exceeding 20 Å, to form a giant CS state. However, the CR process can only take place by a direct transition from this giant CS state to the ground state, and the electronic coupling for this process must be extremely small, owing to the large distance separating the chromophores at the two ends of the system (i.e., $^{+}\mathbf{D}$ and $\mathbf{A}_3^{-}$, Fig. 33). This multichromophore approach should, therefore, allow the efficient formation of long-lived giant CS states. Over the years, some veritable multichromophoric behemoths have been constructed that do, indeed, form long-lived CS states, two spectacular examples of which are shown in Fig. 34. Photoinduced ET in the pentad **37** ultimately leads to the formation of the depicted giant CS state, with a lifetime (in dichloromethane) of 200 μs.[208] Photoinduced ET in the tetrad **38** produces a giant CS state with an even longer lifetime of 380 ms (in a glass at 193 K).[207]

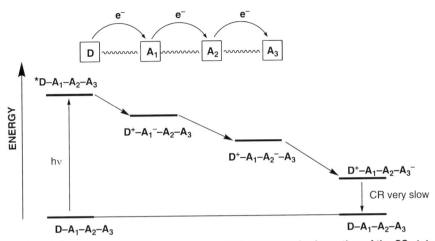

A Series of short, sequential ET hops ensures high efficiency for formation of the CS state

Fig. 33 A schematic example of the formation of a giant photoinduced CS state, $^{+}DA_1A_2A_3^{-}$, in a multichromophoric system using a redox cascade. Note that each ET event in the cascade is exergonic.

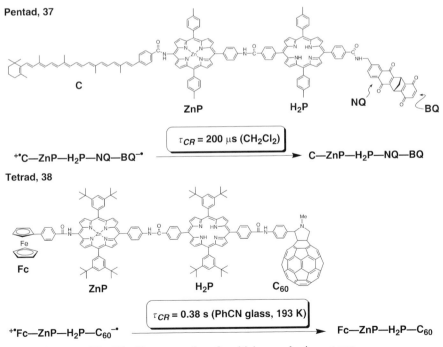

Fig. 34 Two examples of multichromophoric systems.

While this approach is important and is bearing fruit, it suffers from two drawbacks. First, the synthesis of giant multichromophoric systems is often complex and labour-intensive. Secondly, a significant proportion of the photonic excitation energy is generally lost along the charge separation cascade, resulting in a reduced energy conversion yield; consequently, the third requirement, listed above, for an efficient photovoltaic device is not well satisfied. Taking the tetrad **38** as an example, the energy of the excitation photon is 2.1 eV and the driving force for charge recombination from the giant CS state in this molecule (in THF) is 1.07 V.[207] The energy conversion yield is, therefore, 51%.

The second approach to the development of long-lived CS states avoids, in principle, these two drawbacks by resorting to dyads and noting that charge recombination can be slowed down by taking advantage of a difference in electron spin multiplicity between the charge separated state and the ground state. That is, CR from a triplet CS state, to the singlet ground state, is a spin-forbidden process, and should be substantially slower than that from the corresponding singlet CS state, which is spin-allowed. To date, this approach has made use of photoinduced, *non-sensitised*, triplet ET between an electron donor, **D**, and acceptor, **A**, to give a triplet radical ion-pair,^{3}**D**$^{+\cdot}$**A**$^{-\cdot}$, and is well documented.[209-213] Essentially, it involves direct excitation of either **D** or **A**, leading to the first excited singlet state, equation (31), followed by isc, equation (32), to the triplet state, and subsequent charge separation in the triplet manifold, equation (33):

$$\mathbf{D} + h\nu \rightarrow {}^1\mathbf{D}^* \quad \text{or} \quad \mathbf{A} + h\nu \rightarrow {}^1\mathbf{A}^* \tag{31}$$

$$^1\mathbf{D}^* \rightarrow {}^3\mathbf{D}^* \quad \text{or} \quad {}^1\mathbf{A}^* \rightarrow {}^3\mathbf{A}^* \tag{32}$$

$$^3\mathbf{D}^* + \mathbf{A} \rightarrow {}^3\mathbf{D}^{+\cdot}\mathbf{A}^{-\cdot} \quad \text{or} \quad {}^3\mathbf{A}^* + \mathbf{D} \rightarrow {}^3\mathbf{A}^{-\cdot}\mathbf{D}^{+\cdot} \tag{33}$$

$$^1\mathbf{D}^* + \mathbf{A} \rightarrow {}^1\mathbf{D}^{+\cdot}\mathbf{A}^- \quad \text{or} \quad {}^1\mathbf{A}^* + \mathbf{D} \rightarrow {}^1\mathbf{A}^{-\cdot}\mathbf{D}^{+\cdot} \tag{34}$$

The sequence of reactions (31)–(33), illustrated in Fig. 35a for triplet ET occurring from the locally excited triplet acceptor, constitutes a potentially efficient

---

Fig. 35 (a) A schematic describing non-sensitised triplet ET taking place by initial generation of the singlet excited state of the acceptor, followed by rapid isc, to give the locally excited acceptor triplet, followed by ET to give the triplet CS state. This scheme applies equally to either inter-molecular or intra-molecular triplet ET processes. Note that for this reaction scheme to work, the following two conditions should be met: (1) The isc rate, k_{isc}, within the locally excited acceptor should be faster than the singlet charge separation rate, k_{cs}^1. (2) The energy of the locally excited acceptor triplet should be greater than the energy of the triplet CS state; that is, the triplet ET reaction should be exergonic. A similar scheme could be constructed if it were the donor that underwent electronic excitation, and subsequent isc, instead of the acceptor. (b) An example of the generation of a long-lived triplet CS state by non-sensitised triplet ET.[220]

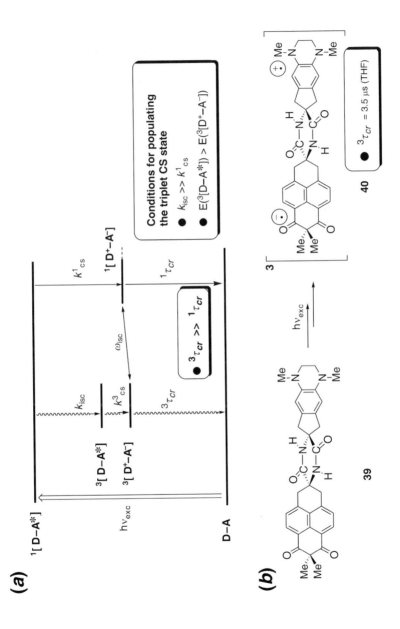

way to achieve *inter-molecular* charge separation, and to retard subsequent charge recombination to the singlet ground state in contact ion pairs (exciplexes). This method has also been applied to *intra-molecular* analogues, in which the two chromophores are covalently linked to flexible polymethylene-type bridges.[214-219] For this non-sensitised triplet ET process to be efficient, two conditions should be met, namely (1) that the isc rate, k_{isc}, for formation of the triplet state from the locally excited singlet state of the acceptor – or donor, whichever may be the case – should be faster than the rate of singlet charge separation, k_{cs}^1, (Fig. 35a), and (2) that the triplet energy of the acceptor (or donor) should be greater than the triplet energy of the CS state; that is, the triplet charge separation process should be exergonic.

For both flexible covalently linked systems and inter-molecular systems, the competition between unwanted spin-allowed singlet charge separation, equation (34), – which has an inherently larger driving force than that for triplet charge separation, equation (33), – and spin-forbidden isc, equation (32), can be effectively avoided if isc takes place at a sufficiently large **D/A** separation, i.e., before the donor and acceptor groups can get close enough for reaction (34) to become important.

Although the efficient formation of the triplet CS state is achievable in flexible covalently linked **D–B–A** dyads having long tethers, these systems are not ideal molecules for producing long-lived CS states because the triplet CS state may decay by rapid isc (ω_{isc} in Fig. 35a) to the singlet CS state which will then undergo speedy spin-allowed CR to the ground state. The magnitude of ω_{isc} is expected to be especially rapid for long tethers (i.e., comprising 10 C–C bonds or more) because, for large **D–A** separations, the exchange interaction, J, approaches, in magnitude, the electron-nuclear hyperfine interactions which are largely responsible for driving the isc process (see Section 8).

It is preferable, therefore to use systems possessing fairly short, rigid bridges. Such bridges should guarantee a sufficiently large magnitude of J to render ω_{isc} small. The requirement of rigidity is a useful one to implement because a bridge which is forced to adopt an all-*trans* configuration will enhance both the triplet ET rate and the magnitude of J. Unfortunately, satisfying these requirements using the non-sensitised triple ET method is difficult to realise in practice because it is limited to systems containing a chromophore, e.g., a quinone, that undergoes extremely rapid isc ($> 10^{10}$ s^{-1}) in order to beat the unwanted singlet charge separation process. Consequently, only a few examples of this method have been reported,[220,221] one of which is shown in Fig. 35b. Local singlet excitation of the naphthalenedione acceptor chromophore in **39** leads to rapid isc to the locally excited triplet state, and then to the triplet CS state, **40**, which was found to have a long lifetime of 3.5 μs.[220]

Hviid and co-workers have recently found a way around this limitation by using the method of intermolecular triplet sensitisation to generate the locally excited triplet state of the donor chromophore in a **D–A** dyad.[222] The sensitiser scheme is described by equations (35)–(38) and is also illustrated graphically in Fig. 36 for the case of triplet sensitisation of a donor chromophore (**S** is the sensitiser molecule). The principal advantage of this method is that the locally excited triplet state of one of the chromophores (in this case, the donor) is populated without invoking its

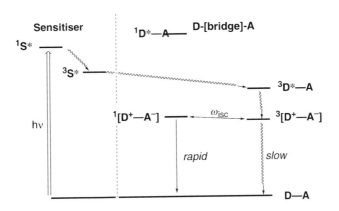

Conditions for sensitized population of the triplet CS state

- $E(^1S*) < E(^1D*—A)$ and $E(^1S*) < E(D—^1A*)$
- $E(^3S*) > E(^3D*—A)$
- $E(^3D*—A) > E(^3D^+—A^-)$

Fig. 36 Schematic energy diagram showing the inter-molecular triplet sensitised population of the locally excited triplet state of the donor group, **D**, in a dyad, **D–A**, by a sensitising molecule, **S**. Note that the singlet energy of **S** must be lower than the singlet energy of either the donor or acceptor, and that the triplet energy of **S** must be greater than that of the donor (or acceptor). Also, for the triplet ET reaction to take place, the locally excited triplet state of the donor must lie above the triplet CS state in energy.

excited singlet state as an intermediate. Consequently, the sensitiser method should be applicable to a broader range of chromophores than is the non-sensitised triplet method.

$$S + h\nu \rightarrow {}^1S^* \tag{35}$$

$$^1S^* \rightarrow {}^3S^* \tag{36}$$

$$^3S^* + D–A \rightarrow {}^3D^*–A + S \tag{37}$$

$$^3D^*–A \rightarrow {}^3(^+D–A^-) \tag{38}$$

The population of the locally exited donor (or acceptor) triplet is achieved simply by excitation of an added external triplet sensitiser molecule whose singlet energy lies below that of both donor and acceptor chromophores, but whose triplet energy lies above that of either the donor or acceptor (Fig. 36).

This method has been applied to the rigid 3-bond dyad, **DNM[3]M**, (Fig. 37), in which the maleate ester group (**M**) serves as the acceptor (the **DCV** acceptor group cannot be used because the energies of the singlet and triplet $^+$**DMN[3]DCV**$^-$ CS states all lie *above* the $^3[^*$**DMN[3]DCV**$]$ locally excited triplet state, see Fig. 29). Benzophenone (**BP**) was used as the sensitiser, and a highly polar solvent

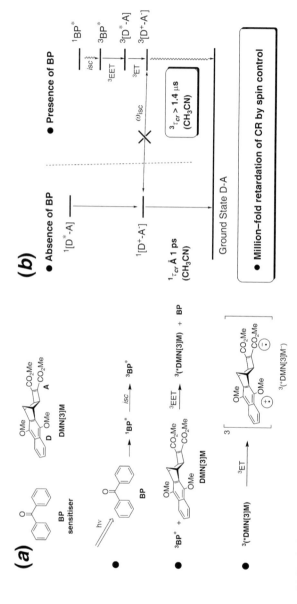

Fig. 37 (a) Depiction of the triplet sensitisation of **DMN[3]M** by benzophenone, to give the locally excited donor triplet state, ^{3,*}**DMN[3]M**, which then undergoes triplet ET to give the triplet CS state, ^{3}($^+$**DMN[3]M** $^-$). (b) Diagram depicting the ET processes taking place in **DMN[3]M** in the absence of benzophenone and in the presence of benzophenone (acetonitrile solvent).

(acetonitrile) was employed, in order to ensure that the triplet ET reaction (^{3}ET; Fig. 37b) is exergonic. In the absence of **BP**, direct excitation of **DMN[3]M** leads to the formation of the singlet CS species, $^1(^+\mathbf{DMN[3]M}^-)$, which undergoes extremely rapid, (sub)picosecond, charge recombination to the ground state. This rapid CR process is in keeping with the strong electronic coupling that exists across the short hydrocarbon bridge in **DNM[3]M**.

Upon triplet sensitisation of the **DMN** chromophore of **DMN[3]M** with **BP**, it was observed (using transient absorption spectroscopy) that triplet EET took place from **BP** to the **DMN** chromophore (^{3}EET; Fig. 37),[179] followed by triplet state charge separation (^{3}ET; Fig. 37). The lifetime of the resulting triplet CS state, $^3(^+\mathbf{DMN[3]M}^-)$, was found to be about 1.4 μs, making it *six orders of magnitude* greater than that for the corresponding singlet CS state!

Whether the slow CR process from the triplet CS state occurs by way of isc to the singlet CS state (ω_{isc}; Fig. 37b), followed by spin-allowed CR from this state, or by a direct, spin-forbidden transition to the ground state is unknown. Recent TD-DFT calculations by Michael Shephard (UNSW) on the singlet and triplet CS states of $^+\mathbf{DMN[3]M}^-$ give a J value of 200 cm^{-1}, which is more than four orders of magnitude greater than either the calculated SOC interaction (0.02 cm^{-1}) between the singlet and triplet CS states or the electron-nuclear hyperfine interaction. These calculations suggest, therefore, that ω_{isc} should be small for the interconversion of the singlet and triplet CS states of **DMN[3]M**, and they underscore the requirement for using "dwarf" dyads, possessing short rigid bridges, for producing long-lived triplet CS states.

The energy conversion yield for photoinduced triplet charge separation in **DMN[3]M** is calculated using the singlet excitation energy of **BP** (3.26 eV) and the driving force for charge recombination (2.62 V).[222] These data give an energy conversion yield of 80%. This is superior to values of about 50% typically obtained for charge separation involving the redox cascade mechanism in giant multi-chromophoric systems. However, we have a long way to go in designing dwarf dyads that will match the millisecond lifetimes observed for the CS states derived from certain multichromophoric systems![206,207]

In summary, we have demonstrated a proof of principle that long-lived triplet CS states in dwarf dyads – that is, dyads possessing short rigid saturated hydrocarbon bridges – may be achieved using the triplet sensitised approach. This sensitisation method offers at least two important advantages over other methods discussed above. These are:

(1) The sensitisation approach should be applicable to a wide range of chromophores, thereby allowing greater flexibility in optimising the three requirements, listed above, for the design of efficient photovoltaic devices.

(2) The sensitisation method will work best for dwarf dyads, rather than for dyads possessing long bridges, because of the requirement that the exchange coupling, J, in the CS state be much larger than both the SOC and hyperfine interactions. Meeting this requirement is a bonus because it means less synthetic work!

10 Symmetry control of ET

There is a third approach to the development of long-lived CS states but it did not work very well in practice. Nevertheless I will briefly discuss it here because it deals with a subject that lies close to the Houkian heart: orbital symmetry control in chemical reactions. In this context, the following simple question is asked: to what degree may the lifetime of a CS state be extended if the CR process is formally forbidden on the grounds of electronic state symmetry? The structural rigidity and symmetry of our norbornylogous bridges, capped at both ends by symmetry-possessing chromophores, make them ideal systems for studying orbital symmetry effects on ET dynamics.[223,224] The results of one such investigation into the effect of electronic state symmetry – or orbital symmetry, to be less precise – on CR rates are summarised in Fig. 38. Assuming that the norbornylogous dyads retain C_s symmetry throughout the CR process, then charge recombination from the singlet CS state of **18(8)**, denoted by $^+$**DMN[8]DCV**$^-$, is formally a symmetry-forbidden process

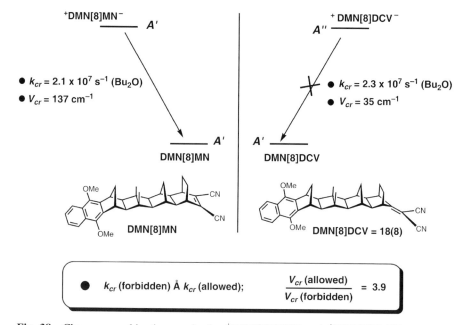

Fig. 38 Charge recombination results for $^+$**DMN[8]MN**$^-$ and $^+$**DMN[8]DCV**$^-$ in butyl ether.[224] The electronic states of $^+$**DMN[8]MN**$^-$ and $^+$**DMN[8]DCV**$^-$ are A' and A'', respectively, and the ground states both have A' symmetry. Thus, assuming that C_s point group symmetry is maintained throughout the CR reaction, then that for $^+$**DMN[8]MN**$^-$ is symmetry-allowed, whereas that for $^+$**DMN[8]DCV**$^-$ is formally symmetry-forbidden – hence the presence of the cross drawn through the reaction arrow. The CR rates are given, together with the electronic coupling matrix elements, V_{cr}. The latter were calculated from the quantum yields and lifetimes for charge transfer fluorescence accompanying charge recombination in $^+$**DMN[8]MN**$^-$ and $^+$**DMN[8]DCV**$^-$.

because the symmetry of the CS state is A'', whereas that of the ground state has A' symmetry. In contrast, the CR process from the singlet CS state of the cognate system, **DMN[8]MN** – in which the **DCV** acceptor is replaced by a maleonitrile (**MN**) acceptor – is formally symmetry-allowed.

Interestingly, the CR rates for these systems in butyl ether are nearly identical – in fact, the symmetry-forbidden process is marginally faster! This result implies that orbital symmetry exerts little control over the dynamics of non-adiabatic ET processes, which may seem a little odd at first sight, considering the profound effect that orbital symmetry has on the relative rates of symmetry-forbidden and symmetry-allowed pericyclic reactions. However, symmetry effects on reactivity are manifested in different ways in pericyclic and ET reactions. In the case of pericyclic reactions, orbital symmetry is manifested by its determining effect on the intended correlation between a given electronic state of the reactant and an electronic state of the product. If the intended correlation is between states having the same level of electronic excitation – i.e., between the ground states of the reactant and product, or between the first excited states of reactant and product, etc. – then the pericyclic reaction is symmetry-allowed. If, on the other hand, the intended correlation involves a product having a higher electronically excited state than the reactant, then the pericyclic reaction is symmetry-forbidden. Clearly, the allowed and forbidden pericyclic processes will, in general, have substantially different activation energies because their intended state correlations are with two different electronic product states having very different energies. Thus, symmetry-forbidden pericyclic reactions are often associated with much higher activation barriers than their allowed counterparts. It should also be noted that symmetry-breaking molecular vibrations will have only a small ameliorating effect on the activation energies of symmetry-forbidden processes.

In contrast, orbital symmetry has only a minor influence on activation energies of ET reactions, because it is only manifested in the electronic coupling term, V_{el} : If the reactant and product states have different electronic state symmetries, then V_{el} is zero and the reaction is formally symmetry-forbidden, whereas if they have the same state symmetry, then V_{el} is non-zero and the reaction is symmetry-allowed. The important point to note is that, for non-adiabatic ET reactions, where V_{el} is very small, often *ca.* 25 cm^{-1}, the allowed and forbidden ET reactions have nearly identical activation energies. This is illustrated in Fig. 39 for allowed and forbidden HT reactions in a radical cation of a complex comprising two ethene molecules. Two geometries for this complex are considered, both possessing C_{2v} symmetry. In Fig. 39a, in which the two ethene moieties lie in mutually perpendicular planes, the HT process is formally forbidden because the reactant and product possess different state symmetries, B_1 and B_2, respectively. The electronic coupling for this reaction is zero. Placement of the two ethene molecules in the same plane (Fig. 39b) now makes the HT process symmetry-allowed and V_{el} is non-zero. It is clear from this figure that, if V_{el} for the allowed HT reaction is small, then both allowed and forbidden HT processes have similar activation energies.

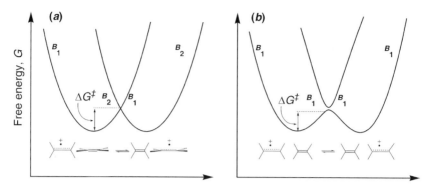

Fig. 39   Illustration of (a) a symmetry-allowed HT process and (b) a symmetry-forbidden HT process. Both reactions take place within a dimeric ethene radical cation complex. Both dimers possess C_{2v} symmetry. For the symmetry-forbidden reaction, (a), the two ethene molecules lie in perpendicular planes; consequently, the reactant and product have different electronic state symmetries, B_1 and B_2, respectively, and V_{el} is therefore zero. For the allowed process, (b), the two ethene groups lie in parallel planes and both reactant and product have identical state symmetries, B_1; thus, V_{el} is non-zero. For non-adiabatic HT, where V_{el} is very small (~ 25 cm^{-1}), the allowed and forbidden processes have nearly identical free energies of activation.

Now, symmetry-forbidden ET processes are readily thwarted because, unlike their pericyclic counterparts, *vibronic coupling* of the reactant and product states with a molecular vibration of the appropriate symmetry – such that the product of the irreducible representations of the reactant and product states, and the molecular vibrational mode contain the totally symmetric representation – will lead to a non-zero value of V_{el} with the consequence that ET will take place, with nearly the same activation free energy as that for a symmetry-allowed process.[225,226] Another way of expressing this point is that symmetry-reducing molecular vibrations take the reaction trajectories for ET into avoided crossing regions of the energy surfaces that are associated with lower molecular symmetries, and where V_{el} is non-zero because, in these regions, the reactant and product states have identical symmetries.

In summary, the above discussion suggests that there is no reason to expect that orbital symmetry effects will have a major influence on ET rates, to the extent that they have on the relative rates of allowed and forbidden pericyclic reactions. Symmetry-breaking vibronic coupling in some symmetry-forbidden ET reactions might even lead to rates that match those for symmetry-allowed processes; indeed, this has been observed for the CR rate in $^+$**DMN[8]DCV**$^-$, relative to $^+$**DMN[8]MN**$^-$. However, because vibronic coupling is generally a weak perturbation, it is likely that the magnitude of V_{el} for a formally symmetry-forbidden ET reaction in a system is smaller than that for the corresponding symmetry-allowed process in a cognate system.

Fortunately, this last point could be confirmed experimentally because CR for both $^+$**DMN[8]MN**$^-$ and $^+$**DMN[8]DCV**$^-$ is accompanied by CT fluorescence.

These CT fluorescence data enabled the unambiguous determination of the electronic coupling, V_{cr}, for CR to be made for each dyad. It was found that V_{cr} for CR in $^+$**DMN[8]MN**$^-$ and $^+$**DMN[8]DCV**$^-$ are 137 and 35 cm^{-1}; respectively.[224] These data show that V_{cr} for the symmetry-allowed CR process is about four times stronger than that for the symmetry-forbidden CR process. Hence, orbital symmetry does have a small, but noticeable, effect on the magnitude of the electronic coupling for CR.

Recently, Garth Jones (UNSW) and his co-workers have described a semi-classical molecular dynamics (MD) model, which incorporates the trajectory surface hopping (TSH) method,[227–229] that shows promise in investigating a number of important problems concerning the effects of molecular vibrations on the dynamics of ET reactions, including formally symmetry-forbidden ones.[230–232] Essentially, the MD-TSH method takes into account the possibility that reaction trajectories for non-adiabatic ET processes may undergo quantum transitions (surface hops) between two potential energy surfaces, as discussed in Section 2 and illustrated in Fig. 5b. Those trajectories which enter the avoided crossing region on the lower potential energy surface, and which experience quantum transitions to the upper surface, take longer to reach the product well and this is reflected in lower frequencies of passage over the reaction barrier or, equivalently, in an electronic transmission coefficient that is less than unity.

As an example of the application of the MD-TSH model to symmetry effects in ET, we employed it to examine the dynamics of thermal hole transfer in the bismethyleneadamantane radical cation, **41**, and the bismethylenehomocubane radical cation, **42** (Fig. 40). Although the adamantane and bishomocubane bridges in these radical cations have the same length, their geometries are such that HT in **41** is formally symmetry-forbidden, because the two active π MOs lie in orthogonal, planes, but is symmetry-allowed in **43**, in which the two π MOs lie in the same plane. Several hundred reaction trajectories were calculated for the two radical cations, using the AM1-CI method to calculate the potential energy surfaces and the forces. As expected, a significantly greater proportion of trajectories for **41** underwent multiple quantum transitions than for **42**, reflecting the smaller magnitude of the electronic coupling in the symmetry-forbidden process. The frequencies of passage over the reaction barrier to form product were calculated to be 9.6×10^{12} and 1.6×10^{13} s^{-1} for **41** and **42**, respectively. The frequency of passage for the symmetry-allowed HT in **42** is, therefore, 1.6 times greater than that for the symmetry-forbidden process in **41**. The average value of the electronic coupling, $\langle V_{el} \rangle$, for HT in **41** and **42** is 80 and 1050 cm^{-1}, respectively. Thus, $\langle V_{el} \rangle$ is about 13 times larger for the symmetry-allowed HT process in **42** than for the symmetry-forbidden process in **41**. These trends are similar to those found experimentally for charge recombination in $^+$**DMN[8]MN**$^-$ and $^+$**DMN[8]DCY**$^-$ (Fig. 38); i.e., there is little difference between the rates of the allowed and forbidden ET reactions, but there is a moderately significant difference in the magnitudes of the electronic coupling matrix elements. The reason for the non-zero value of $\langle V_{el}K \rangle$ for the formally symmetry-forbidden HT reaction in **41**, and the nearly equal

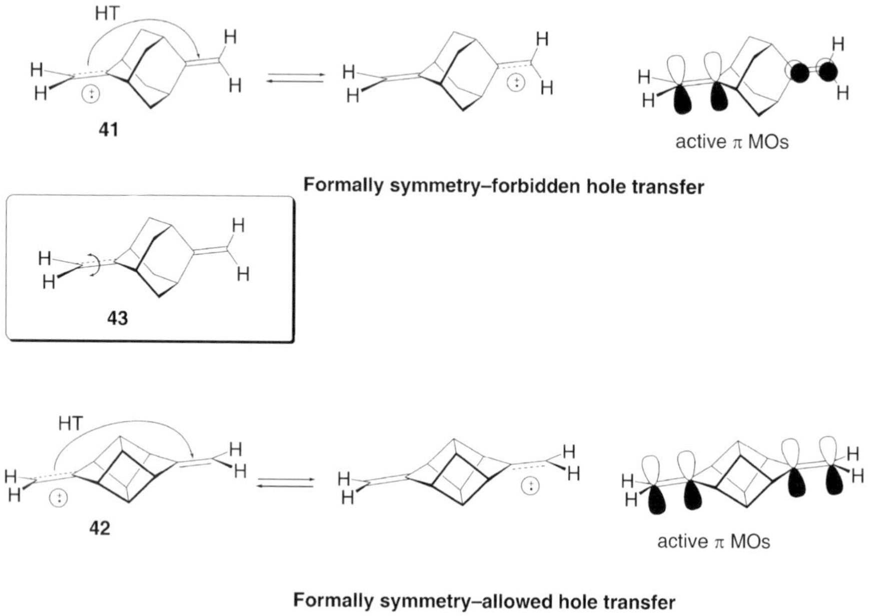

Formally symmetry–forbidden hole transfer

Formally symmetry–allowed hole transfer

Fig. 40   Thermal hole transfer processes in bismethyleneadamantane radical cation, **41**, and bismethylenebishomocubane radical cation, **42**; the former process is formally symmetry-forbidden, because the two active π MOs are orthogonal to each other, whereas the latter is symmetry-allowed, because the two active π MOs lie in the same plane. The inset, **43**, depicts the major symmetry-breaking vibration, a torsional mode, which facilitates HT in **41**.

frequencies of passage of the forbidden and allowed HT process for **41** and **42**, respectively, was traced to a symmetry-breaking torsional vibration about the terminal methylene group which is associated with the one-electron π-bond, e.g., **43** (Fig. 40, inset).[230]

Thus, molecular vibrations play a crucial role in ET processes, and MD calculations involving quantum transitions, such as our MD-TSH method, offer considerable promise as a powerful tool for investigating their role in ET reactions.

11   Concluding remarks

The past 20 years have witnessed enormous progress in our understanding of the character of long-range, non-adiabatic, ET processes and this has been achieved, in no small part, by the study of ET in covalently linked multichromophoric systems. Using an elegant combination of synthesis, photophysical measurements and computational quantum chemical calculations, the scope and significance of the superexchange mechanism in ET reactions have been delineated. For example, it is now known that ET, mediated by superexchange coupling through saturated hydrocarbon bridges, can take place rapidly over distances exceeding 12 Å.

ET dynamics may also be modulated in a predictable manner, for example, by altering the bridge configuration, by the presence of strong external electric fields,[233,234] and by using different bridge, such as hydrocarbon bridges, proteins, peptides and DNA helices. The variation of the distance dependence of ET dynamics with the degree of unsaturation in bridges is now well-understood, and the transition from the coherent superexchange-mediated ET mechanism to various types of incoherent *electron transport* mechanisms (solitons, polarons, charge-hopping, etc.) in conjugated bridges, has been observed. Electron conduction has also been observed in single molecules, including saturated,[235] conjugated,[236] and DNA[237,238] molecules.

These significant findings form the basis of a set of design principles for the construction of molecular photovoltaic cells, and other nanoscale electronic devices, in which the control of both the rate and directionality of ET processes is an essential requirement. The successful construction of an artificial light-driven proton pump, based on principles of long-range ET processes illustrates the promise of this approach.[239]

It is impractical to synthesise nanoscale devices based solely on covalently linked redox systems and it is reassuring to learn that ET is also strongly mediated by the superexchange mechanism in non-covalently linked H-bonded networks[240-249] and by solvent molecules,[202,250-264] thereby opening the way for the construction of photovoltaic supramolecular assemblies.

While our knowledge of the fundamental properties of ET has been greatly increased, the goal of applying that knowledge to the successful rational design, construction, and operation of molecular electronic devices remains elusive – but not, I suspect, for long. Hopefully, chemists from various disciplines will be inspired to pursue the many important challenges – of both fundamental and applied character – that remain in the ET field.[3]

Acknowledgements

I am indebted to the outstanding and essential contributions made by a large number of graduate students and postdoctoral research fellows who, over the years at UNSW, have synthesised so many varieties of wonderful molecules that have led to revelations in the ET field. Particular acknowledgements for synthetic work are made to the late Evangelo Cotsaris and to Drs Anna M. Oliver, Nicholas Head, M. G. Ranasinghe, Kate Jolliffe, Nigel Lokan, Anna Golka-Bartlett, James Lawson, Daniel Rothenfluh, Andrew Black, Christoph Mecker, and Steven J. Langford. Computational quantum chemistry plays a crucial role in our investigations and three brilliant theoretical chemistry colleagues have made seminal contributions. They are Dr Garth Jones, Michael Shephard, and Stephen Wong. On the photophysical side, my collaboration with Professors Jan W. Verhoeven (Amsterdam) and John M. Warman (Delft), ongoing since 1985, has been extraordinarily successful and productive, and I thank them for their generosity, friendship, and for sharing their brilliant insights with me. In a similar vein, I am equally indebted to Professors Kenneth D. Jordan (Pittsburgh), Barry K. Carpenter (Cornell), and David N. Beratan (Pittsburgh) for many fruitful collaborations on the theory side. I also thank Professors Kenneth P. Ghiggino (Melbourne), David H. Waldeck (Pittsburgh), Hanns Fischer (Zürich), Siegfried Schneider (Erlangen-Nürnberg), Fabian Gerson (Basel), and Yves Rubin (UCLA) for many exciting experimental collaborations. Continued, generous support from the Australian Research Council, and its predecessors, over the years is gratefully acknowledged, as is the award of an ARC Senior Research Fellowship. Support from the NSW State and National Super-computing Facilities, ac3 and APAC, respectively, is acknowledged.

References

1. Gray, H.B. and Winkler, J.R. (2001). In *Electron Transfer in Chemistry*, Balzani, V. (ed.), vol. 3, p. 3, Part 1, Chapter 1. Wiley-VCH, Weinheim
2. Jortner, J. and Bixon, M. (eds) (1999). In *Electron Transfer—From Isolated Molecules to Biological Molecules*, Parts 1 and 2. Wiley Interscience, New York
3. Balzani, V. (ed.), (2001). In *Electron Transfer in Chemistry*. vols. 1–5. Wiley-VCH, Weinheim
4. Hush, N.S. (1985). *Coord. Chem. Rev.* **64**, 135
5. Marcus, R.A. and Sutin, N. (1985). *Biochim. Biophys. Acta* **811**, 265
6. Bixon, M. and Jortner, J. (1999). *Adv. Chem. Phys.* **106**, 35
7. Newton, M.D. (2001). In *Electron Transfer in Chemistry*, Balzani, V. (ed.), vol. 1, p. 3, Part 1, Chapter 1. Wiley-VCH, Weinheim
8. Bolton, J.R. and Archer, M.D. (1991). ACS Advances in Chemistry Series. In *Electron Transfer in Inorganic, Organic, and Biological Systems*, Bolton, J.R., Mataga, N. and McLendon, G. (eds), vol. 228, p. 7, Washington
9. Miller, J.R., Calcaterra, L.T. and Closs, G.L. (1984). *J. Am. Chem. Soc.* **106**, 3047
10. Deisenhofer, J. and Michel, H. (1989). *Angew. Chem. Int. Ed. Engl.* **28**, 829
11. Boxer, S.G., Goldstein, R.A., Lockhart, D.J., Middendorf, T.R. and Takiff, L. (1989). *J. Phys. Chem.* **93**, 8280
12. Thompson, M.A. and Zerner, M.C. (1990). *J. Am. Chem. Soc.* **112**, 7828
13. Osuka, A., Marumo, S., Mataga, N., Taniguchi, S., Okada, T., Yamazaki, I., Nishimura, Y., Ohno, T. and Nozaki, K. (1996). *J. Am. Chem. Soc.* **118**, 155
14. Paddon-Row, M.N. and Jordan, K.D. (1988). In *Modern Models of Bonding and Delocalization*, Liebman, J.F. and Greenberg, A. (eds), vol. 6, p. 115. VCH Publishers, New York
15. Miller, J.R. (1987). *New J. Chem.* **11**, 83
16. Miller, J.R., Beitz, J.V. and Huddleston, R.K. (1984). *J. Am. Chem. Soc.* **106**, 5057
17. Roth, S. (1995). *One-Dimensional Metals*. VCH, Weinheim
18. Felts, A.K., Pollard, W.T. and Friesner, R.A. (1995). *J. Phys. Chem.* **99**, 2929
19. Davis, W.B., Wasielewski, M.R., Ratner, M.A., Mujica, V. and Nitzan, A. (1997). *J. Phys. Chem. A* **101**, 6158
20. Okada, A., Chernyak, V. and Mukamel, S. (1998). *J. Phys. Chem. A* **102**, 1241
21. Nitzan, A. (2001). *J. Phys. Chem. A* **105**, 2677
22. Nitzan, A. (2001). *Ann. Rev. Phys. Chem.* **52**, 681
23. Lehmann, J., Kohler, S., Hanggi, P. and Nitzan, A. (2002). *Phys. Rev. Lett.* **8822**, 8305
24. Tolbert, L.M. (1992). *Acc. Chem. Res.* **25**, 561

25. Tolbert, L.M. and Zhao, X.D. (1997). *J. Am. Chem. Soc.* **119**, 3253
26. Jortner, J., Bixon, M., Langenbacher, T. and Michel-Beyerle, M.E. (1998). *Proc. Natl Acad. Sci. USA* **95**, 12759
27. Bixon, M., Giese, B., Wessely, S., Langenbacher, T., Michel-Beyerle, M.E. and Jortner, J. (1999). *Proc. Natl Acad. Sci. USA* **96**, 11713
28. Giese, B. (2000). *Acc. Chem. Res.* **33**, 631
29. Giese, B., Amaudrut, J., Kohler, A.K., Spormann, M. and Wessely, S. (2001). *Nature* **412**, 318
30. Giese, B., Spichty, M. and Wessely, S. (2001). *Pure Appl. Chem.* **730**, 449
31. Lewis, F.D., Letsinger, R.L. and Wasielewski, M.R. (2001). *Acc. Chem. Res.* **34**, 159
32. Kramers, H.A. (1934). *Physica* **1**, 182
33. Anderson, P.W. (1950). *Phys. Rev.* **79**, 350
34. McConnell, H.M. (1961). *J. Chem. Phys.* **35**, 508
35. Salem, L. (1982). *Electrons in Chemical Reactions: First Principles*. Wiley Interscience, New York
36. Hoffmann, R. (1971). *Acc. Chem. Res.* **4**, 1
37. Paddon-Row, M.N. (1982). *Acc. Chem. Res.* **15**, 245
38. Balaji, V., Ng, L., Jordan, K.D., Paddon-Row, M.N. and Patney, H.K. (1987). *J. Am. Chem. Soc.* **109**, 6957
39. Jordan, K.D. and Paddon-Row, M.N. (1992). *Chem. Rev.* **92**, 395
40. Williams, R.M., Koeberg, M., Lawson, J.M., An, Y.Z., Rubin, Y., Paddon-Row, M.N. and Verhoeven, J.W. (1996). *J. Org. Chem.* **61**, 5055
41. Paddon-Row, M.N. (2001). In *Electron Transfer in Chemistry*, Balzani, V. (ed.), vol. 3, p. 179, Part 2, Chapter 1. Wiley-VCH, Weinheim
42. Larsson, S. (1981). *J. Am. Chem. Soc.* **103**, 4034
43. Larsson, S. (1983). *J. Chem. Soc. Faraday Trans.* **2**, 1375
44. Larsson, S. and Braga, M. (1993). *Chem. Phys.* **176**, 367
45. Newton, M.D. (1991). *Chem. Rev.* **91**, 767
46. Wasielewski, M.R. (1992). *Chem. Rev.* **92**, 435
47. Weissman, S.I. (1958). *J. Am. Chem. Soc.* **80**, 6462
48. Voevodskii, V.V., Solodovnikov, S.P. and Chibrikin, V.N. (1959). *Doklady Akad. Nauk S.S.S.R.* **129**, 1082
49. Shimada, K., Moshuk, G., Connor, H.D., Caluwe, P. and Szwarc, M. (1972). *Chem. Phys. Lett.* **14**, 396
50. Shimada, K. and Szwarc, M. (1974). *Chem. Phys. Lett.* **28**, 540
51. Gerson, F., Huber, W., Martin, W.B., Caluwe, J.P., Pepper, T. and Szwarc, M. (1984). *Helv. Chim. Acta* **67**, 416
52. Pasman, P., Verhoeven, J.W. and de Boer, T.J. (1976). *Tetrahedron* **37**, 2827
53. Pasman, P., Verhoeven, J.W. and de Boer, T.J. (1977). *Tetrahedron Lett.* **17**, 207
54. Pasman, P., Rob, F. and Verhoeven, J.W. (1982). *J. Am. Chem. Soc.* **104**, 5127
55. Verhoeven, J.W. (1999). *Adv. Chem. Phys.* **106**, 603
56. Verhoeven, J.W., Dirkx, I.P. and de Boer, T.J. (1966). *Tetrahedron Lett.* **25**, 4399
57. Verhoeven, J.W., Dirkx, I.P. and de Boer, T.J. (1969). *Tetrahedron* **25**, 339
58. Verhoeven, J.W., Dirkx, I.P. and de Boer, T.J. (1969). *Tetrahedron* **25**, 4037
59. Pasman, P., Verhoeven, J.W. and de Boer, T.J. (1978). *Chem. Phys. Lett.* **59**, 381
60. Pasman, P., Mes, G.F., Koper, N.W. and Verhoeven, J.W. (1985). *J. Am. Chem. Soc.* **107**, 5839
61. Crawford, M.K., Wang, Y. and Eisenthal, K.B. (1981). *Chem. Phys. Lett.* **79**, 529
62. Wang, Y., Crawford, M.K. and Eisenthal, K.B. (1982). *J. Am. Chem. Soc.* **104**, 529
63. Mataga, N. (1984). *Pure Appl. Chem.* **56**, 1255
64. Mataga, N. (1993). *Pure Appl. Chem.* **65**, 1605

65. Yang, N.C., Neoh, S.B., Naito, T., Ng, L.K., Chernoff, D.A. and McDonals, D.B. (1980). *J. Am. Chem. Soc.* **102**, 2806
66. van Dantzig, N.A., Shou, H.S., Alfano, J.C., Yang, N.C. and Levy, D.H. (1994). *J. Chem. Phys.* **100**, 7068
67. Zhang, S.L., Lang, M.J., Goodman, S., Durnell, C., Fidlar, V., Fleming, G.R. and Yang, N.C. (1996). *J. Am. Chem. Soc.* **118**, 9042
68. Yang, N.C., Zhang, S.L., Lang, M.J., Goodman, S., Durnell, C., Fleming, G.R., Carrell, H.L. and Garavito, R.M. (1999). *Adv. Chem. Phys.* **106**, 645
69. Paddon-Row, M.N. and Hartcher, R. (1980). *J. Am. Chem. Soc.* **102**, 662
70. Paddon-Row, M.N. and Hartcher, R. (1980). *J. Am. Chem. Soc.* **102**, 671
71. McGourty, J.L., Blough, N.V. and Hoffman, B.M. (1983). *J. Am. Chem. Soc.* **105**, 4470
72. Gray, H.B. (1986). *Chem. Soc. Rev.* **15**, 17
73. Isied, S.S. (1991). In *Electron Transfer in Inorganic, Organic, and Biological Systems*, Bolton, J.R., Mataga, N. and McLendon, G. (eds), vol. 228, p. 229. American Chemical Society, Washington, DC
74. Isied, S.S., Ogawa, M.Y. and Wishart, J.F. (1992). *Chem. Rev.* **92**, 381
75. Langen, R., Colon, J.L., Casimiro, D.R., Karpishin, T.B., Winkler, J.R. and Gray, H.B. (1996). *J. Biol. Inorg. Chem.* **1**, 221
76. Fleming, G.R., Martin, J.L. and Breton, J. (1988). *Nature (London)* **333**, 190
77. Martin, J.L., Breton, J., Hoff, A.J., Migus, A. and Antonetti, A. (1986). *Proc. Natl Acad. Sci. USA* **83**, 957
78. Holten, D., Windsor, M.W., Parson, W.W. and Thornber, J.P. (1978). *Biochim. Biophys. Acta* **501**, 112
79. Paddon-Row, M.N., Patney, H.K., Brown, R.S. and Houk, K.N. (1981). *J. Am. Chem. Soc.* **103**, 5575
80. Balaji, V., Jordan, K.D., Burrow, P.D., Paddon-Row, M.N. and Patney, H.K. (1982). *J. Am. Chem. Soc.* **104**, 6849
81. Jørgensen, F.S., Paddon-Row, M.N. and Patney, H.K. (1983). *J. Chem. Soc. Chem. Commun.* 573
82. Paddon-Row, M.N., Patney, H.K., Peel, J.B. and Willett, G.D. (1984). *J. Chem. Soc. Chem. Commun.* 564
83. Jordan, K.D. and Paddon-Row, M.N. (1992). *J. Phys. Chem.* **96**, 1188
84. Shephard, M.J., Paddon-Row, M.N. and Jordan, K.D. (1993). *Chem. Phys.* **176**, 289
85. Paddon-Row, M.N., Shephard, M.J. and Jordan, K.D. (1993). *J. Phys. Chem.* **97**, 1743
86. Koopmans, T. (1934). *Physica* **1**, 104
87. Falcetta, M.F., Jordan, K.D., McMurry, J.E. and Paddon-Row, M.N. (1990). *J. Am. Chem. Soc.* **112**, 579
88. Reed, A.E., Curtiss, L.A. and Weinhold, F. (1988). *Chem. Rev.* **88**, 899
89. Jordan, K.D., Nachtigallova, D. and Paddon-Row, M.N. (1997), In *Modern Electronic Structure Theory and Applications in Organic Chemistry*, Davidson, E.R. (ed.), p. 257. World Scientific Publishing Co, Singapore
90. Calcaterra, L.T., Closs, G.L. and Miller, J.R. (1983). *J. Am. Chem. Soc.* **105**, 670
91. Paddon-Row, M.N., Cotsaris, E. and Patney, H.K. (1986). *Tetrahedron* **42**, 1779
92. Oevering, H., Paddon-Row, M.N., Heppener, H., Oliver, A.M., Cotsaris, E., Verhoeven, J.W. and Hush, N.S. (1987). *J. Am. Chem. Soc.* **109**, 3258
93. Paddon-Row, M.N., Oliver, A.M., Symons, M.C.R., Cotsaris, E., Wong, S.S. and Verhoeven, J.W. (1989). In *Photoconversion Processes for Energy and Chemicals*, Hall, D.O. and Grassi, G. (eds), vol. 5, p. 79. Elsevier Applied Science, London
94. Antolovich, M., Oliver, A.M. and Paddon-Row, M.N. (1989). *J. Chem. Soc. Perkin Trans. 2*, 783
95. Craig, D.C., Lawson, J.M., Oliver, A.M. and Paddon-Row, M.N. (1990). *J. Chem. Soc. Perkin Trans. 1*, 3305

96. Oliver, A.M. and Paddon-Row, M.N. (1990). *J. Chem. Soc. Perkin Trans. 1*, 1145
97. Paddon-Row, M.N. and Verhoeven, J.W. (1991). *New J. Chem.* **15**, 107
98. Paddon-Row, M.N. (1994). *Acc. Chem. Res.* **27**, 18
99. Warman, J.M., de Haas, M.P., Verhoeven, J.W. and Paddon-Row, M.N. (1999). *Adv. Chem. Phys.* **106**, 571
100. Craig, D.C. and Paddon-Row, M.N. (1987). *Aust. J. Chem.* **40**, 1951
101. Paddon-Row, M.N., Oliver, A.M., Warman, J.M., Smit, K.J., de Haas, M.P., Oevering, H. and Verhoeven, J.W. (1988). *J. Phys. Chem.* **92**, 6958
102. Hush, N.S., Paddon-Row, M.N., Cotsaris, E., Oevering, H., Verhoeven, J.W. and Heppener, M. (1985). *Chem. Phys. Lett.* **117**, 8
103. Warman, J.M., de Haas, M.P., Paddon-Row, M.N., Cotsaris, E., Hush, N.S., Oevering, H. and Verhoeven, J.W. (1986). *Nature* **320**, 615
104. Warman, J.M., Smit, K.J., de Haas, M.P., Jonker, S.A., Paddon-Row, M.N., Oliver, A.M., Kroon, J., Oevering, H. and Verhoeven, J.W. (1991). *J. Phys. Chem.* **95**, 1979
105. Seischab, M., Lodenkemper, T., Stockmann, A., Schneider, S., Koeberg, M., Roest, M.R., Verhoeven, J.W., Lawson, J.M. and Paddon-Row, M.N. (2000). *Phys. Chem. Chem. Phys.* **2**, 1889
106. Kroon, J., Verhoeven, J.W., Paddon-Row, M.N. and Oliver, A.M. (1991). *Angew. Chem. Int. Ed. Engl.* **30**, 1358
107. Kroon, J., Oliver, A.M., Paddon-Row, M.N. and Verhoeven, J.W. (1988). *Recl. Trav. Chim. Pays-Bas* **107**, 509
108. Oliver, A.M., Craig, D.C., Paddon-Row, M.N., Kroon, J. and Verhoeven, J.W. (1988). *Chem. Phys. Lett.* **150**, 366
109. Wasielewski, M.R., Niemczyk, M.P., Johnson, D.G., Svec, W.A. and Minsek, D.W. (1989). *Tetrahedron* **45**, 4785
110. Oevering, H., Verhoeven, J.W., Paddon-Row, M.N. and Warman, J.M. (1989). *Tetrahedron* **45**, 4751
111. Penfield, K.W., Miller, J.R., Paddon-Row, M.N., Cotsaris, E., Oliver, A.M. and Hush, N.S. (1987). *J. Am. Chem. Soc.* **109**, 5061
112. Hush, N.S. (1968). *Electrochim. Acta* **13**, 1005
113. Hush, N.S. (1967). *Prog. Inorg. Chem.* **8**, 391
114. Clayton, A.H.A., Ghiggino, K.P., Wilson, G.J., Keyte, P.J. and Paddon-Row, M.N. (1992). *Chem. Phys. Lett.* **195**, 249
115. Stein, C.A., Lewis, N.A. and Seitz, G. (1982). *J. Am. Chem. Soc.* **104**, 2596
116. Johnson, M.D., Miller, J.R., Green, N.S. and Closs, G.L. (1989). *J. Phys. Chem.* **93**, 1173
117. Closs, G.L. and Miller, J.R. (1988). *Science* **240**, 440
118. Closs, G.L., Calcaterra, L.T., Green, N.J., Penfield, K.W. and Miller, J.R. (1986). *J. Phys. Chem.* **90**, 3673
119. Yonemoto, E.H., Saupe, G.B., Schmehl, R.H., Hubig, S.M., Riley, R.L., Iverson, B.L. and Mallouk, T.E. (1994). *J. Am. Chem. Soc.* **116**, 4786
120. Park, J.W., Lee, B.A. and Lee, S.Y. (1998). *J. Phys. Chem. B* **102**, 8209
121. Wold, D.J., Haag, R., Rampi, M.A. and Frisbie, C.D. (2002). *J. Phys. Chem. B* **106**, 2813
122. Shephard, M.J. and Paddon-Row, M.N. (1995). *J. Phys. Chem.* **99**, 17497
123. Paddon-Row, M.N. and Shephard, M.J. (1997). *J. Am. Chem. Soc.* **119**, 5355
124. Shephard, M.J. and Paddon-Row, M.N. (1999). *Chem. Phys. Lett.* **301**, 281
125. Naleway, C.A., Curtiss, L.A. and Miller, J.R. (1991). *J. Phys. Chem.* **95**, 8434
126. Liang, C. and Newton, M.D. (1992). *J. Phys. Chem.* **96**, 2855
127. Liang, C. and Newton, M.D. (1993). *J. Phys. Chem.* **97**, 3199
128. Koga, N., Sameshima, K. and Morokuma, K. (1993). *J. Phys. Chem.* **97**, 13117
129. Paddon-Row, M.N., Wong, S.S. and Jordan, K.D. (1990). *J. Am. Chem. Soc.* **112**, 1710
130. Paddon-Row, M.N., Wong, S.S. and Jordan, K.D. (1990). *J. Chem. Soc. Perkin Trans. 2*, 417

131. Paddon-Row, M.N. and Jordan, K.D. (1993). *J. Am. Chem. Soc.* **115**, 2952
132. Beratan, D.N. and Hopfield, J.J. (1984). *J. Am. Chem. Soc.* **106**, 1584
133. Paddon-Row, M.N., Wong, S.S. and Jordan, K.D. (1990). *J. Chem. Soc. Perkin Trans.* 2, 425
134. Beratan, D.N. (1986). *J. Am. Chem. Soc.* **108**, 4321
135. Beratan, D.N., Onuchic, J.N. and Hopfield, J.J. (1985). *J. Chem. Phys.* **83**, 5326
136. Onuchic, J.N. and Beratan, D.N. (1987). *J. Am. Chem. Soc.* **109**, 6771
137. Paddon-Row, M.N., Shephard, M.J. and Jordan, K.D. (1993). *J. Am. Chem. Soc.* **115**, 3312
138. Curtiss, L.A., Naleway, C.A. and Miller, J.R. (1995). *J. Phys. Chem.* **99**, 1182
139. Curtiss, L.A., Naleway, C.A. and Miller, J.R. (1993). *Chem. Phys.* **176**, 387
140. Curtiss, L.A., Naleway, C.A. and Miller, J.R. (1993). *J. Phys. Chem.* **97**, 4050
141. Heilbronner, E. and Schmelzer, A. (1975). *Helv. Chim. Acta* **58**, 936
142. Weber, K., Hockett, L. and Creager, S. (1997). *J. Phys. Chem. B* **101**, 8286
143. Creager, S., Yu, C.J., Bamdad, C., O'Connor, S., MacLean, T., Lam, E., Chong, Y., Olsen, G.T., Luo, J.Y., Gozin, M. and Kayyem, J.F. (1999). *J. Am. Chem. Soc.* **121**, 1059
144. Bakkers, E., Marsman, A.W., Jenneskens, L.W. and Vanmaekelbergh, D. (2000). *Angew. Chem. Int. Ed.* **39**, 2297
145. Bakkers, E., Roest, A.L., Marsman, A.W., Jenneskens, L.W., de Jong-van Steensel, L.I., Kelly, J.J. and Vanmaekelbergh, D. (2000). *J. Phys. Chem. B* **104**, 7266
146. Sachs, S.B., Dudek, S.P., Hsung, R.P., Sita, L.R., Smalley, J.F., Newton, M.D., Feldberg, S.W. and Chidsey, C.E.D. (1997). *J. Am. Chem. Soc.* **119**, 10563
147. Sikes, H.D., Smalley, J.F., Dudek, S.P., Cook, A.R., Newton, M.D., Chidsey, C.E.D. and Feldberg, S.W. (2001). *Science* **291**, 1519
148. Davis, W.B., Svec, W.A., Ratner, M.A. and Wasielewski, M.R. (1998). *Nature* **396**, 60
149. Winkler, J.R. and Gray, H.B. (1992). *Chem. Rev.* **92**, 369
150. Moser, C.C., Keske, J.M., Warncke, K., Farid, R.S. and Dutton, P.L. (1992). *Nature* **355**, 796
151. Beratan, D.N., Onuchic, J.N. and Hopfield, J.J. (1987). *J. Chem. Phys.* **86**, 4488
152. Beratan, D.N., Betts, J.N. and Onuchic, J.N. (1991). *Science* **252**, 1285
153. Regan, J.J. and Onuchic, J.N. (1999). *Adv. Chem. Phys.* **107**, 497
154. Skourtis, S.S. and Beratan, D.N. (1999). *Adv. Chem. Phys.* **106**, 377
155. Onuchic, J.N., Beratan, D.N., Winkler, J.R. and Gray, H.B. (1992). *Ann. Rev. Biophys. Biomol. Struct.* **21**, 349
156. Barbara, P.F. (1999). *Adv. Chem. Phys.* **107**, 647
157. Lewis, F.D. (2001). In *Electron Transfer in Chemistry*, Balzani, V. (ed.), vol. 3, p. 105, Part 1, Chapter 5. Wiley-VCH, Weinheim
158. Meggers, E., Michel-Beyerle, M.E. and Giese, B. (1998). *J. Am. Chem. Soc.* **120**, 12950
159. Giese, B., Wessely, S., Spormann, M., Lindemann, U., Meggers, E. and Michel-Beyerle, M.E. (1999). *Angew. Chem. Int. Ed.* **38**, 996
160. Meggers, E., Kusch, D., Spichty, M., Wille, U. and Giese, B. (1998). *Angew. Chem. Int. Ed.* **37**, 460
161. Giese, B. and Wessely, S. (2001). *Chem. Commun.* 2108
162. Steenken, S. and Jovanovic, S.V. (1997). *J. Am. Chem. Soc.* **119**, 617
163. Giese, B. and Spichty, M. (2000). *ChemPhysChem* **1**, 195
164. Segal, D., Nitzan, A., Davis, W.B., Wasielewski, M.R. and Ratner, M.A. (2000). *J. Phys. Chem. B* **104**, 3817
165. Grozema, F.C., Berlin, Y.A. and Siebbeles, L.D.A. (2000). *J. Am. Chem. Soc.* **122**, 10903
166. Aflatooni, K., Gallup, G.A. and Burrow, P.D. (1998). *J. Phys. Chem. A* **102**, 6205
167. Lewis, F.D., Wu, T.F., Liu, X.Y., Letsinger, R.L., Greenfield, S.R., Miller, S.E. and Wasielewski, M.R. (2000). *J. Am. Chem. Soc.* **122**, 2889

168. Lewis, F.D., Wu, T.F., Zhang, Y.F., Letsinger, R.L., Greenfield, S.R. and Wasielewski, M.R. (1997). *Science* **277**, 673
169. Kelley, S.O. and Barton, J.K. (1999). *Science* **283**, 375
170. Harriman, A. (1999). *Angew. Chem. Int. Ed.* **38**, 945
171. Brun, A.M. and Harriman, A. (1992). *J. Am. Chem. Soc.* **114**, 3656
172. Olson, E.J.C., Hu, D.H., Hormann, A. and Barbara, P.F. (1997). *J. Phys. Chem. B* **101**, 299
173. Fukui, K. and Tanaka, K. (1998). *Angew. Chem. Int. Ed.* **37**, 158
174. Fukui, K., Tanaka, K., Fujitsuka, M., Watanabe, A. and Ito, O. (1999). *J. Photochem. Photobiol. B Biol.* **50**, 18
175. Hess, S., Gotz, M., Davis, W.B. and Michel-Beyerle, M.E. (2001). *J. Am. Chem. Soc.* **123**, 10046
176. Hess, S., Davis, W.B., Voityuk, A.A., Rosch, N., Michel-Beyerle, M.E., Ernsting, N.P., Kovalenko, S.A. and Lustres, J.L.P. (2002). *ChemPhysChem* **3**, 452
177. Davis, W.B., Hess, S., Naydenova, I., Haselsberger, R., Ogrodnik, A., Newton, M.D. and Michel-Beyerle, M.E. (2002). *J. Am. Chem. Soc.* **124**, 2422
178. Turro, N.J. and Barton, J.K. (1998). *J. Biol. Inorg. Chem.* **3**, 201
179. Roest, M.R., Oliver, A.M., Paddon-Row, M.N. and Verhoeven, J.W. (1997). *J. Phys. Chem. A* **101**, 4867
180. Paddon-Row, M.N. and Shephard, M.J. (2002). *J. Phys. Chem. A* **106**, 2935
181. Wegner, M., Fischer, H., Grosse, S., Vieth, H.M., Oliver, A.M. and Paddon-Row, M.N. (2001). *Chem. Phys.* **264**, 341
182. Salem, L. and Rowland, C. (1972). *Angew. Chem. Int. Ed. Engl.* **92**, 11
183. Rajca, A. (1994). *Chem. Rev.* **94**, 871
184. Ogrodnik, A., Remy-Richter, N. and Michel-Beyerle, M.E. (1987). *Chem. Phys. Lett.* **135**, 567
185. Bixon, M., Jortner, J., Michel-Beyerle, M.E., Ogrodnik, A. and Lersch, W. (1987). *Chem. Phys. Lett.* **140**, 626
186. Bixon, M., Jortner, J. and Michel-Beyerle, M.E. (1993). *Z. Phys. Chem.* **180**, 193
187. Volk, M., Haberle, T., Feick, R., Ogrodnik, A. and Michel-Beyerle, M.E. (1993). *J. Phys. Chem.* **97**, 9831
188. Sekiguchi, S., Kobori, Y., Akiyama, K. and Tero-Kubota, S. (1998). *J. Am. Chem. Soc.* **120**, 1325
189. Kobori, Y., Sekiguchi, S., Akiyama, K. and Tero-Kubota, S. (1999). *J. Phys. Chem. A* **103**, 5416
190. Kobori, Y., Akiyama, K. and Tero-Kubota, S. (2000). *J. Chem. Phys.* **113**, 465
191. Forbes, M.D.E. (1993). *J. Phys. Chem.* **97**, 3396
192. Forbes, M.D.E., Closs, G.L., Calle, P. and Gautam, P. (1993). *J. Phys. Chem.* **97**, 3384
193. Forbes, M.D.E., Ball, J.D. and Avdievich, N.I. (1996). *J. Am. Chem. Soc.* **118**, 4707
194. Closs, G.L., Piotrowiak, P., MacInnes, J.M. and Fleming, G.R. (1988). *J. Am. Chem. Soc.* **110**, 2652
195. Closs, G.L., Forbes, M.D.E. and Piotrowiak, P. (1992). *J. Am. Chem. Soc.* **114**, 3285
196. Forbes, M.D.E. and Ruberu, S.R. (1993). *J. Phys. Chem.* **97**, 13223
197. Forbes, M.D.E., Ruberu, S.R. and Dukes, K.E. (1994). *J. Am. Chem. Soc.* **116**, 7299
198. Tsentalovich, Y.P., Morozova, O.B., Avdievich, N.I., Ananchenko, G.S., Yurkovskaya, A.V., Ball, J.D. and Forbes, M.D.E. (1997). *J. Phys. Chem.* **101**, 8809
199. Wegner, M., Fischer, H., Koeberg, M., Verhoeven, J.W., Oliver, A.M. and Paddon-Row, M.N. (1999). *Chem. Phys.* **242**, 227
200. Warman, J.M., Smit, K.J., Jonker, S.A., Verhoeven, J.W., Oevering, H., Kroon, J., Paddon-Row, M.N. and Oliver, A.M. (1993). *Chem. Phys.* **170**, 359
201. Gust, D., Moore, T.A. and Moore, A.L. (1993). *Acc. Chem. Res.* **26**, 198

202. Lawson, J.M., Paddon-Row, M.N., Schuddeboom, W., Warman, J.M., Clayton, A.H.A. and Ghiggino, K.P. (1993). *J. Phys. Chem.* **97**, 13099
203. Imahori, H. and Sakata, Y. (1997). *Adv. Mater* **9**, 537
204. Jolliffe, K.A., Langford, S.J., Ranasinghe, M.G., Shephard, M.J. and Paddon-Row, M.N. (1999). *J. Org. Chem.* **64**, 1238
205. Jolliffe, K.A., Langford, S.J., Oliver, A.M., Shephard, M.J. and Paddon-Row, M.N. (1999). *Chem. Eur. J.* **5**, 2518–2530
206. Gust, D., Moore, T.A. and Moore, A.L. (2001). In *Electron Transfer In Chemistry*, Balzani, V. (ed.), vol. 3, p. 272, Part 2, Chapter 1. Wiley-VCH, Weinheim
207. Imahori, H., Guldi, D.M., Tamaki, K., Yoshida, Y., Luo, C.P., Sakata, Y. and Fukuzumi, S. (2001). *J. Am. Chem. Soc.* **123**, 6617
208. Gust, D., Moore, T.A., Moore, A.L., Macpherson, A.N., Lopez, A., Degraziano, J.M., Gouni, I., Bittersmann, E., Seely, G.R., Gao, F., Nieman, R.A., Ma, X.C.C., Demanche, L.J., Hung, S.C., Luttrull, D.K., Lee, S.J. and Kerrigan, P.K. (1993). *J. Am. Chem. Soc.* **115**, 11141
209. Becker, H.G.O., Lehnmann, T. and Zieba, J. (1989). *J. Prakt. Chem.* **331**, 805
210. Kapinus, E.I. and Aleksankina, M.M. (1990). *Russ. J. Phys. Chem.* **64**, 1413
211. Whitten, D.G. (1978). *Rev. Chem. Intermed.* **2**, 107
212. Yasuike, M., Shima, M., Koseki, K., Yamaoka, T., Sakuragi, M. and Ichimura, K. (1992). *J. Photochem. Photobiol. A* **64**, 115
213. Levin, P.P., Pluzhnikov, P.F. and Kuzmin, V.A. (1988). *Chem. Phys. Lett.* **147**, 283
214. Weller, A., Staerk, H. and Treichel, R. (1984). *Faraday Discuss. Chem. Ser.* **78**, 271
215. Staerk, H., Busmann, H.D., Kühnle, W. and Weller, A. (1989). *Chem. Phys. Lett.* **155**, 603
216. Staerk, H., Busmann, H.D., Kühnle, W. and Treichel, R. (1991). *J. Phys. Chem.* **95**, 1906
217. Busmann, H.D., Staerk, H. and Weller, A. (1989). *J. Chem. Phys.* **91**, 4098
218. Shafirovich, V.Y., Batova, E.E. and Levin, P.P. (1993). *Z. Phys. Chem.* **182**, 254
219. Smit, K.J. and Warman, J.M. (1988). *J. Lumin.* **42**, 149
220. Anglos, D., Bindra, V. and Kuki, A. (1994). *J. Chem. Soc. Chem. Commun.* 213
221. van Dijk, S.I., Groen, C.P., Hartl, F., Brouwer, A.M. and Verhoeven, J.W. (1996). *J. Am. Chem. Soc.* **118**, 8425
222. Hviid, L., Brouwer, A.M., Paddon-Row, M.N. and Verhoeven, J.W. (2001). *ChemPhysChem* **2**, 232
223. Zeng, Y. and Zimmt, M.B. (1991). *J. Am. Chem. Soc.* **113**, 5107
224. Oliver, A.M., Paddon-Row, M.N., Kroon, J. and Verhoeven, J.W. (1992). *Chem. Phys. Lett.* **191**, 371
225. Reimers, J.R., Hush, N.S., Sammeth, D.M. and Callis, P.R. (1990). *Chem. Phys. Lett.* **169**, 622
226. Reimers, J.R. and Hush, N.S. (1990). *Chem. Phys.* **146**, 105
227. Tully, J.C. (1990). *J. Chem. Phys.* **93**, 1061
228. Chapman, S. (1992). *Adv. Chem. Phys.* **82**, 423
229. Hack, M.D. and Truhlar, D.G. (2000). *J. Phys. Chem. A* **104**, 7917
230. Jones, G.A., Paddon-Row, M.N., Carpenter, B.K. and Piotrowiak, P. (2002). *J. Phys. Chem. A* **106**, 5011
231. Jones, G.A., Carpenter, B.K. and Paddon-Row, M.N. (1998). *J. Am. Chem. Soc.* **120**, 5499
232. Jones, G.A., Carpenter, B.K. and Paddon-Row, M.N. (1999). *J. Am. Chem. Soc.* **121**, 11171
233. Galoppini, E. and Fox, M.A. (1996). *J. Am. Chem. Soc.* **118**, 2299
234. Fox, M.A. and Galoppini, E. (1997). *J. Am. Chem. Soc.* **119**, 5277
235. Cui, X.D., Primak, A., Zarate, X., Tomfohr, J., Sankey, O.F., Moore, A.L., Moore, T.A., Gust, D., Harris, G. and Lindsay, S.M. (2001). *Science* **294**, 571

236. Donhauser, Z.J., Mantooth, B.A., Kelly, K.F., Bumm, L.A., Monnell, J.D., Stapleton, J.J., Price, D.W., Rawlett, A.M., Allara, D.L., Tour, J.M. and Weiss, P.S. (2001). *Science* **292**, 2303
237. Fink, H.W. and Schonenberger, C. (1999). *Nature* **398**, 407
238. Porath, D., Bezryadin, A., de Vries, S. and Dekker, C. (2000). *Nature* **403**, 635
239. Steinberg-Yfrach, G., Liddell, P.A., Hung, S.C., Moore, A.L., Gust, D. and Moore, T.A. (1997). *Nature* **385**, 239
240. Chang, C.J., Brown, J.D.K., Chang, M.C.Y., Baker, E.A. and Nocera, D.G. (2001). In *Electron Transfer In Chemistry*, Balzani, V. (ed.), vol. 3, p. 409, Part 2, Chapter 4. Wiley-VCH, Weinheim
241. de Rege, P.J.F., Williams, S.A. and Therien, M.J. (1995). *Science* **269**, 1409
242. Zaleski, J.M., Chang, C.K., Leroi, G.E., Cukier, R.I. and Nocera, D.G. (1992). *J. Am. Chem. Soc.* **114**, 3564
243. Turro, C., Chang, C.K., Leroi, G.E., Cukier, R.I. and Nocera, D.G. (1992). *J. Am. Chem. Soc.* **114**, 4013
244. Roberts, J.A., Kirby, J.P. and Nocera, D.G. (1995). *J. Am. Chem. Soc.* **117**, 8051
245. Berman, A., Izraeli, E.S., Levanon, H., Wang, B. and Sessler, J.L. (1995). *J. Am. Chem. Soc.* **117**, 8252
246. Hayashi, T. and Ogoshi, H. (1997). *Chem. Soc. Rev.* **26**, 355
247. Ward, M.D. (1997). *Chem. Soc. Rev.* **26**, 365
248. Kirby, J.P., Roberts, J.A. and Nocera, D.G. (1997). *J. Am. Chem. Soc.* **119**, 9230
249. Berg, A., Shuali, Z., Asano-Someda, M., Levanon, H., Fuhs, M. and Mobius, K. (1999). *J. Am. Chem. Soc.* **121**, 7433
250. Roest, M.R., Lawson, J.M., Paddon-Row, M.N. and Verhoeven, J.W. (1994). *Chem. Phys. Lett.* **230**, 536
251. Roest, M.R., Verhoeven, J.W., Schuddeboom, W., Warman, J.M., Lawson, J.M. and Paddon-Row, M.N. (1996). *J. Am. Chem. Soc.* **118**, 1762
252. Jolliffe, K.A., Bell, T.D.M., Ghiggino, K.P., Langford, S.J. and Paddon-Row, M.N. (1998). *Angew. Chem. Int. Ed.* **37**, 916
253. Bell, T.D.M., Jolliffe, K.A., Ghiggino, K.P., Oliver, A.M., Shephard, M.J., Langford, S.J. and Paddon-Row, M.N. (2000). *J. Am. Chem. Soc.* **122**, 10661
254. Lokan, N.R., Paddon-Row, M.N., Koeberg, M. and Verhoeven, J.W. (2000). *J. Am. Chem. Soc.* **122**, 5075
255. Napper, A.M., Read, I., Waldeck, D.H., Head, N.J., Oliver, A.M. and Paddon-Row, M.N. (2000). *J. Am. Chem. Soc.* **122**, 5220
256. Koeberg, M., de Groot, M., Verhoeven, J.W., Lokan, N.R., Shephard, M.J. and Paddon-Row, M.N. (2001). *J. Phys. Chem. A* **105**, 3417
257. Napper, A.M., Head, N.J., Oliver, A.M., Shephard, M.J., Paddon-Row, M.N., Read, I. and Waldeck, D.H. (2002). *J. Am. Chem. Soc.* **124**, 10171
258. Goes, M., de Groot, M., Koeberg, M., Verhoeven, J.W., Lokan, N.R., Shephard, M.J. and Paddon-Row, M.N. (2002). *J. Phys. Chem. A* **106**, 2129
259. Kumar, K., Lin, Z., Waldeck, D.H. and Zimmt, M.B. (1996). *J. Am. Chem. Soc.* **118**, 243
260. Han, H. and Zimmt, M.B. (1998). *J. Am. Chem. Soc.* **120**, 8001
261. Read, I., Napper, A., Kaplan, R., Zimmt, M.B. and Waldeck, D.H. (1999). *J. Am. Chem. Soc.* **121**, 10976
262. Read, I., Napper, A., Zimmt, M.B. and Waldeck, D.H. (2000). *J. Phys. Chem. A* **104**, 9385
263. Kaplan, R.W., Napper, A.M., Waldeck, D.H. and Zimmt, M.B. (2000). *J. Am. Chem. Soc.* **122**, 12039
264. Napper, A.M., Read, I., Kaplan, R., Zimmt, M.B. and Waldeck, D.H. (2002). *J. Phys. Chem. A* **106**, 5288

Structure and reactivity of hydrocarbon radical cations

Olaf Wiest,* Jonas Oxgaard and Nicolas J. Saettel

Department of Chemistry and Biochemistry, University of Notre Dame, Notre Dame, Indiana, USA

1 Introduction

Radical cations are interesting intermediates which can be formed by electron transfer from many organic substrates. Their unique structure and reactivity made them attractive topics for a large number of studies in classical physical organic chemistry, often using the techniques of radiation chemistry such as matrix isolation and ESR spectroscopy.[1] More recently, organic chemists looking for new reactions with new selectivities incorporated radical cations into complex reaction sequences[2] or used them for activation of unactivated carbon–hydrogen bonds.[3] Finally, various radical ions are increasingly under investigation as reactive intermediates in a variety of biologically important reaction mechanisms.[4]

These developments have in turn renewed the interest in understanding the structure and reactivity of radical cations. Modern computational chemistry methods, especially density functional methods, as well as the continued exponential increase in hardware performance provided improved tools for a detailed analysis of these interesting species. At the same time, the unique problems of the computational treatment of radical cations as well as the direct and indirect observation of these short-lived species continue to pose new challenges for the development of new theoretical and experimental methods.

---

* Corresponding author.

87

ADVANCES IN PHYSICAL ORGANIC CHEMISTRY
VOLUME 38 ISSN 0065-3160   DOI 10.1016/S0065-3160(03)38002-5

The purpose of this review is to discuss some recent computational studies of radical cations in the context of qualitative concepts of classical physical organic chemistry. In particular, we will demonstrate how such basic, well-understood concepts such as conjugation and electronic state or even more fundamental notions of structure, bonding, and mechanism can lead to new and interesting effects in radical cation chemistry, which are quite different than what is usually expected in the chemistry of neutral compounds. We will also discuss how these effects need to be taken into consideration to understand the chemistry of radical cations. This relatively broad scope means that this review will necessarily be limited to a focused discussion rather than a comprehensive review of the different aspects of radical cation chemistry. Thus, we will concentrate on computational results from our own laboratory, and will discuss experimental data only in the context of the calculational data. A number of recent reviews[5] and book chapters[6] provide much more detail on aspects that cannot be covered in this limited contribution.

2 Computational treatment of radical cations

The calculation of radical cations is still a challenge for modern electronic structure methods.[7] As with any computational study, the three main issues are the necessary basis set, the computational method used for the accurate treatment of electron correlation, and the choice of the model system. While in the case of radical cations, the standard 6-31G* basis set is usually considered to be sufficient for most geometry optimizations and triple zeta basis sets such as the 6-311 + G** basis set are adequate even for highly correlated MO methods, the choice of the computational method for the treatment of electron correlation is less clear. Many of the radical cations discussed here are inherently two-configuration systems, but MCSCF or CASSCF methods are less frequently applied. This is because of the difficulties in choosing the active space since there is, for example, little difference between formally single and double bonds, as will be discussed later. In addition, dynamic correlation, which is not considered in CASSCF calculation, is also significant due to the typically large number of closely spaced electronic states. Perturbation theory approaches such as MP2 have been frequently used, but are very sensitive to spin contamination of the underlying Hartree–Fock wavefunction. When comparing two stationary points on a hypersurface, even small changes in the $\langle S^2 \rangle$ between the two species can lead to substantial changes in the relative energies, and may distort the computed hypersurface by introducing spurious minima. As a result of several documented cases of this effect,[8] the use of MP2 calculations for the calculation of radical cations is not generally recommended.

Two approaches to electron correlation that are widely used today for the studies of organic radical cations are Coupled Cluster (CC) calculations or the similar, but not identical, Quadratic Configuration Interaction (QCI) method with single and double excitations, often followed by CCSD(T) or QCISD(T) single point calculations with a larger basis set. These methods suffer to a much lesser extent from the problems of spin contamination of the underlying Hartree–Fock wavefunction discussed above and give results in excellent agreement with experimental data. However, the high computational demands and unfavorable scaling factor typically limits the application of these methods to small model systems with typically seven non-hydrogen atoms or less.

Since substituents can significantly alter the electronic structure of a system, a more realistic chemical model including these substituents often needs to be studied. This is only feasible using a computationally more efficient, but possibly less accurate method that is validated against the high-level reference CC or QCI calculations. Density functional theory (DFT) calculations, especially hybrid DFT methods such as the B3LYP or BHandH functionals, have been very successful in the study of radical cations. Typically, the geometries, relative energies and even hyperfine coupling constants are in very good agreement with QCISD(T)//QCISD calculations and available experimental data, respectively. Nevertheless, these methods have a bias towards delocalized structures that originates in the non-exact exchange correlation of DFT.[9] This leads, for example, in some cases to a qualitatively incorrect description of reactions.[10] Since conjugation and delocalization are important factors in the chemistry of radical cations, special care needs to be taken in such systems to ensure the validity of a computed result. Nevertheless, hybrid density functional calculations have become the most widely used computational method for the study of radical cations. Thus, all results discussed in this review were obtained at the B3LYP/6-31G* level of theory unless specified otherwise.

3   Symmetry and electronic states

The presence of an unpaired electron in radical cations has significant consequences for symmetric compounds. This becomes clear if one considers the transformation of an open shell symmetric compound into another compound of a different symmetry. In such cases, the symmetry of the singly occupied molecular orbital (SOMO) determines the overall electronic states of the reactant and the product. If the two electronic states do not correlate, i.e., do not share a common symmetry element, a symmetry-preserving pathway from reactant to product is not possible. Any adiabatic reaction leading from the reactant to the product therefore has to involve the loss of symmetry. This problem obviously does not occur for the case of closed-shell molecules, where all orbitals are doubly occupied, leading to a common electronic A_1 state for all molecules.

In addition, reactants or transition states can be subject to a first-order Jahn–Teller effect if symmetry leads to a degenerate state. According to the original definition, "the nuclear configuration of any nonlinear polyatomic system in a degenerate electronic state is unstable with respect to nuclear displacements that lower the symmetry and remove the degeneracy".[11] Therefore, many symmetric radical cations undergo first-order Jahn–Teller distortion. Furthermore, even in cases of nondegenerate states, the vibronic coupling of the radical cation ground state with an low-lying excited state of different symmetry may lead to symmetry-lowering distortions occuring spontaneously or via very low barriers. Such so-called second-order Jahn–Teller effects[12] are in fact quite common in radical cations. This is due to the relatively small HOMO–SOMO and SOMO–LUMO gaps that make a variety of such interactions possible.

A case where both effects are important is the ring opening of the cyclobutene radical cation $\mathbf{1}^{\cdot+}$.[13] This reaction was studied experimentally by Bally and coworkers, who found that, in contrast to the thermal reaction, only the *trans*-1,3-butadiene radical cation, *trans*-$\mathbf{2}^{\cdot+}$, is formed in a matrix isolation experiment.[14,15] This surprising result can be explained by considering the electronic states of the species involved in the reaction, which are shown in Fig. 1.

The cyclobutere radical cation $\mathbf{1}^{\cdot+}$ has a C_{2v} symmetry and a $^2B_1$ ground state. Following a conrotatory pathway leads to the first excited state of *cis*-$\mathbf{2}^{\cdot+}$, which also has a C_{2v} symmetry and a $^2B_1$ state. Conversely, the $^2A_2$ ground state of *cis*-$\mathbf{2}^{\cdot+}$ correlates to the second excited state of $\mathbf{1}^{\cdot+}$. Thus, a direct, symmetry conserving reaction of the group state of $\mathbf{1}^{\cdot+}$ to the ground state of *cis*-$\mathbf{2}^{\cdot+}$ is not possible.

In contrast, *trans*-$\mathbf{2}^{\cdot+}$ has a C_{2h} symmetry and a ^2B_g ground state, therefore in principle allowing a symmetry preserving pathway to correlate it with the ground state of $\mathbf{1}^{\cdot+}$. The corresponding C_2-symmetric transition structure has, at the QCISD(T)//QCISD level of theory and using a 6-31G* basis set, an activation energy of 23.4 kcal/mol. However, a frequency analysis of this transition structure

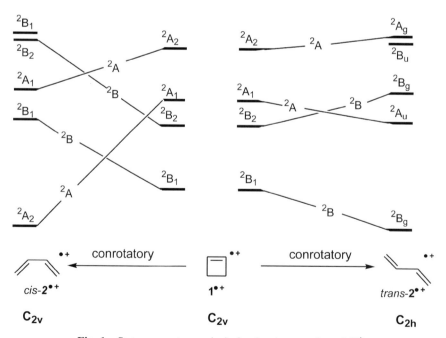

Fig. 1 State symmetry analysis for the ring opening of $\mathbf{1}^{\cdot+}$.

reveals it as a second-order saddle point, i.e., it has two imaginary frequencies. It corresponds to a second-order Jahn–Teller distortion due to vibronic coupling with an excited state that gets into close proximity of the ground state at this geometry. Hence, this transformation also entails a loss of symmetry.

Based on this analysis, there are three remaining pathways possible for the electrocyclic ring opening of $1^{\cdot+}$, which are together with their activation energies at the QCISD(T)/6-31G*//QCISD/6-31G* activation energies summarized in Fig. 2. A concerted mechanism leading to cis-$2^{\cdot+}$ can proceed via an asymmetric transition structure $3^{\cdot+}$ with an activation energy of 18.1 kcal/mol. Alternatively, a stepwise pathway involving the cyclopropyl carbinyl radical cation intermediate $4^{\cdot+}$ has been located using MP2 calculation.[16] Higher levels of theory demonstrate that this is, at least for the parent system, not a stationary point on the hypersurface. Electron donating substituents can stabilize $4^{\cdot+}$ and make this a minimum on the hypersurface, but even in those cases, a concerted pathway is preferred.[17] Finally, Bally and coworkers located a second concerted pathway involving an asymmetric transition structure $5^{\cdot+}$, which leads to $trans$-$2^{\cdot+}$ with an activation energy of 20.9 kcal/mol, only 2.8 kcal/mol higher in energy than $3^{\cdot+}$.[18] As will be discussed later, this energetic ordering depends on the computational method as well as the model system used and is, for example, reversed by including solvent effects through a cavity-type solvent model.[19]

Similar results are also obtained for the next higher homologue of the reaction, the ring closing reaction of the 1,3,5-hexatriene radical cation $6^{\cdot+}$ to give the 1,3-cyclohexadiene radical cation $7^{\cdot+}$. Even though $6^{\cdot+}$ is spectroscopically well

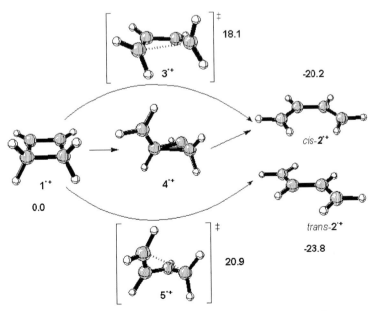

Fig. 2 Pathways for the electrocyclic ring opening of $1^{\cdot+}$.

characterized[20] and the ring closing reaction is calculated to be exothermic by 23.8 kcal/mol,[21] it has not been observed experimentally for the parent $6^{\cdot+}$.[22] Instead, the photochemically induced ring opening of $7^{\cdot+}$ has been described in detail in the literature.[23] Thus, a computational study of the mechanisms has to address the question under which circumstances this reaction occurs (Fig. 3).

Similar to the case of $1^{\cdot+}$, there are in principle two possible concerted pathways, one involving the C_2 symmetric transition structure $8^{\cdot+}$ and an unsymmetric transition structure $9^{\cdot+}$. Although $8^{\cdot+}$ is a true transition structure, which is not subject to a Jahn–Teller distortion and has only one negative eigenfrequency, it is not part of the reaction pathway, because it not only has an activation barrier of 27.8 kcal/mol relative to $6^{\cdot+}$, but also there is no adiabatic connection between the ^2B ground state of $6^{\cdot+}$ and the ^2A state of $7^{\cdot+}$. Thus, the unsymmetric pathway involving $9^{\cdot+}$ is the only viable option for a concerted pathway with an activation energy of 16 kcal/mol.

Competing with this pathway is the stepwise ring closure of the cis,cis,trans1,3,5-hexatriene radical cation $11^{\cdot+}$ to give the [3.1.0]-bicyclohexene radical cation $13^{\cdot+}$. While the transition structure $12^{\cdot+}$ for this reaction is only 1 kcal/mol lower in energy than the one for the concerted pathway, $11^{\cdot+}$ is 5.4 kcal/mol lower in energy. The activation energy for the reaction from $11^{\cdot+}$ to $13^{\cdot+}$ is with 20.4 kcal/mol substantially higher than the one for the concerted pathway. The high activation barrier for rotation around the formal single C_2–C_3 bond prevents the isomerization from $6^{\cdot+}$ to $11^{\cdot+}$ at the radical cation stage. Rather, a back electron transfer, followed by rapid isomerization at the stage of the neutral molecule and subsequent reoxidation to the radical cation is the preferred pathway for the isomerization of the different rotamers of $6^{\cdot+}$.[21] Since 11 is 5.9 kcal/mol lower in energy than 6, a substantially larger population of 1,3,5-hexatriene would exist as the conformer 11 in the equilibrium, which is then preferentially oxidized. Ring opening of $13^{\cdot+}$ and subsequent hydrogen shift then leads to the final product of the reaction $7^{\cdot+}$.

The existence of different rotamers of $6^{\cdot+}$ and the relatively high barriers separating them is also the reason why the ring closing reaction is not observed experimentally. The most stable isomer for both neutral 1,3,5-hexatriene and its

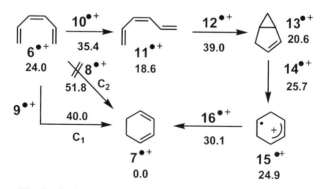

Fig. 3 Pathways for the Electrocyclic ring opening of $6^{\cdot+}$.

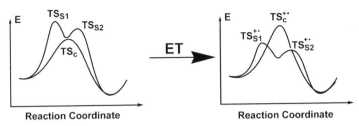

Fig. 4 Effect of ET catalysis on the relative energies of concerted and stepwise pathways.

radical cation is the *trans,trans,trans* isomer, which is 11.5 kcal/mol more stable than **6**$^{\cdot+}$, but cannot undergo ring closure to **7**$^{\cdot+}$. Through the electron transfer–isomerization–electron transfer sequence described above, **6**$^{\cdot+}$ or **11**$^{\cdot+}$ would rapidly isomerize to give the unreactive *trans,trans,trans* isomer. Only 1,3,5-hexatriene radical cations that are conformationally restricted, such as the one studied by Barkow and Grützmacher,[22] will therefore undergo cyclization.

It is interesting to note that the role of symmetry in these formally pericyclic reactions of radical cations is very different to the one in neutral reactions. There, the Woodward–Hoffmann rules and the resulting aromatic character of the transition states favor symmetric pathways. In the case of the radical cation reactions, limitations placed on symmetric pathway by the requirement for electronic state correlation and Jahn–Teller distortions often disfavor symmetric, concerted reaction pathways. Instead, many pathways were found to resemble the ones for biradical, stepwise pathways of neutral pericyclic reactions. This can be rationalized by considering the effect of ET on the relative energies of competing stepwise and concerted pathway as indicated in Fig. 4. In a normal pericyclic reaction, the symmetric transition state **TS$_c$** is stabilized by aromaticity and is lower in energy than the stepwise pathway, which typically requires the initial disruption of a bond in the first transition state **TS$_{S1}$**. The corresponding radical cation that is obtained after electron transfer is not aromatic and thus not stabilized. Even in the absence of a Jahn–Teller effect, the activation energy of the symmetric pathway would thus be raised relative to the competing stepwise reaction. Conversely, the initial breaking of a single or double bond is facilitated in the case of the radical cation, lowering the activation energy of that pathway. Even though these effects might not necessarily be large enough to enforce a stepwise reaction, they will typically make the concerted and stepwise pathways energetically competitive. In many cases, solvent and substituent effects will be large enough to lead to change in the reaction mechanism.[17,19]

4   Conjugation

Conjugation, the stabilization of chemical species through interactions with π-systems, is a fundamental concept in organic chemistry.[24] Every student of elementary organic chemistry is expected to be familiar with the concepts of conjugation and delocalization in order to understand the chemistry of allylic or aromatic systems. Consequently, a significant portion of any course in the field is devoted to a discussion of the energetic importance of conjugation relative to other influences such as hybridization or sterics. At the same time, the same effects are of current research interest to explain phenomena such as organic conductivity or fluorescence.[25]

The energetic effects of conjugation are largest when empty or half-empty p-orbitals interact with a π-system. Typical examples include allyl cations or allyl radicals, respectively. In these cases, the allylic stabilization was estimated to be ~20 kcal/mol.[26] In comparison, the effect on neutral, closed-shell molecules is relatively small. The conjugative effect on the rotation of 1,3-butadiene **2** is, for example, with 3 kcal/mol much smaller.

The large stabilization of conjugative interactions in radical cations has unexpected effects on their structure and reactivity. One example is the *cis/trans* isomerization of the 1,3-butadiene radical cation $2^{\cdot +}$.[27] It can be seen from the bond distances shown in Fig. 5 that in both *cis*-$2^{\cdot +}$ and *trans*-$2^{\cdot +}$, the radical cation is delocalized over the entire molecule. This leads to a very short C_2–C_3 bond length of 1.42 and 1.41 Å in *cis* and *trans*-$2^{\cdot +}$, respectively.[13] There is basically no distinction between the formal single and double bonds. Bally and coworkers[8a,b] already noted that the transition structure for the isomerization of *cis* and *trans* cannot be located using the B3LYP methodology, but that increasing the amount of

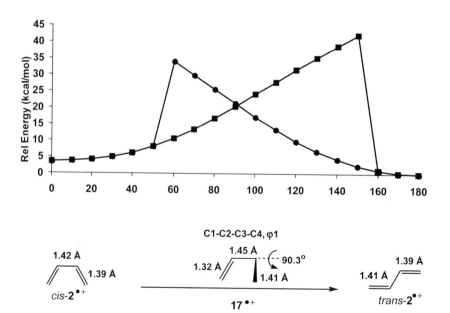

Fig. 5 B3LYP/6-31G* rotational profile for the isomerization of $2^{\cdot +}$ starting from *cis*-$2^{\cdot +}$ (squares) and *trans*-$2^{\cdot +}$ (circles).

Hartree–Fock exchange in the BHandH functional allows the localization of the transition structure $\mathbf{17}^{\cdot+}$, where a localization of spin and charge in one part of the molecule took place, as can be seen from the large differences in bond lengths.

Although Wiberg pointed out substantial differences between theoretical and experimental values, the stabilizing effect of the conjugative interaction is often evaluated by rotation around the central carbon–carbon bond for a variety of systems,[28–33] including the butadiene radical cation,[27] and derivatives thereof. However, B3LYP/6-31G* calculations of the C_2–C_3 rotation in $\mathbf{2}^{\cdot+}$, while allowing the localization of spin and charge in an unsymmetric structure, yielded the surprising plot shown in Fig. 5. After reaching the transition structure region with a C_1–C_2–C_3–C_4 dihedral angle φ_1 of approximately 90°, the calculated energy continues to rise until almost reaching a planar structure. At this point, a process other than rotation around the C_2–C_3 bond leads to a rapid drop in energy. In addition, the rotational profiles computed starting from *cis* and *trans* $\mathbf{2}^{\cdot+}$ are very different. Validation of the results by QCISD/6-31G* calculations, shown in Fig. 6, confirm that this is not a computational artifact related to the previously discussed inability of the B3LYP method to locate the transition structure for this rotation.[9,10] Examination of the geometries obtained for the crossing point of the two scans reveals the origin of this behavior. Upon rotation around the C_2–C_3 bond, these two carbon centers undergo significant rehybridization from a H–C–C–H dihedral angle φ_2 of 12.1° in *cis*-$\mathbf{2}^{\cdot+}$ to the structure shown in Fig. 6, where this angle is 26.7°. The driving force for this rehybridization is the need to maintain conjugation with the double bond in order to stabilize the localized spin and charge in the transition structure region. At the true transitions structure, which can be localized by either QCISD or BH and H calculations, C_2 and C_3 are planarized.

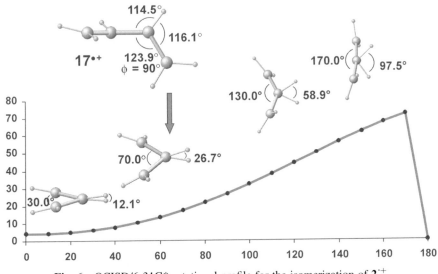

Fig. 6 QCISD/6-31G* rotational profile for the isomerization of $\mathbf{2}^{\cdot+}$.

It is clear from these results that the tendency to maintain conjugation as long as possible during the rotation leads to a mechanism for the *cis/trans* isomerization of $2^{\cdot+}$ that is quite different than the one for neutral **2**. While in the latter case, the φ_1 is sufficient as a reaction coordinate, the rotation of $2^{\cdot+}$ needs two dihedral angles to be described adequately: one that describes the torsion of the carbon framework and one that describes the rehybridization of the central carbons such as φ_2. As can be seen in Fig. 7, only φ_1 changes at the beginning of the rotation, while φ_2 remains constant. Upon further rotation, the central carbons are increasingly pyramidalized as indicated by changes in φ_2. At a φ_1 of approximately 85°, the pyramidalized carbons invert via the true transition structure of the isomerization and eventually leading to a planar structure.

The strong interactions along the pathway have a number of practical consequences. Despite the considerable flexibility of $2^{\cdot+}$ around the broad minimum shown in Fig. 7, the activation energy for rotation around the formal C_2–C_3 double bond $2^{\cdot+}$ is at 28.1 kcal/mol,[27] much higher than the ~ 3 kcal/mol rotational barrier for **2**. This means that short polyene radical cations do not undergo rotation and are locked in one conformation. The easiest way for rotation is usually an electron transfer to regenerate the neutral species, which can then undergo rapid rotation for the reasons already discussed for the case of $6^{\cdot+}$. For longer polyene chains or for substituted cases, the localized structures will get more stabilized since other means of stabilizing spin and charge are available. Thus, the rehybridization will be less important for these cases and almost negligible for polyenes with highly stabilizing substituents. Thus, the activation energies for the isomerization of the 2,3-dihydroxy-1,3-butadiene radical cation $18^{\cdot+}$ and 2,3-disilyl-1,3-butadiene

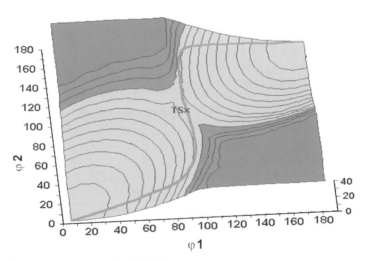

Fig. 7   3D representation of the B3LYP hypersurface of the isomerization of $2^{\cdot+}$. Each contour line corresponds to 3 kcal/mol. The reaction pathway shown in Fig. 5 is indicated in bold.

radical cation **19**$^{\cdot+}$ are at 12.8 and 8.8 kcal/mol, respectively, much lower than for isomerization of **2**$^{\cdot+}$.[34] These substituents provide sufficient stabilization in the transition state so that it can be easily located even using B3LYP calculations.

5 Bonding

The mechanical model of the chemical bond as a spring with a given force constant and a typical equilibrium length that can be modified only in a relatively narrow range and then ruptures by passing through a transition state has been an important concept in chemistry. From the interpretation of infrared spectra to the parametrization of molecular mechanics force fields, this concept has been widely used and an intuitive knowledge of typical bond lengths is an important part of a chemist's toolbox. In order to explore the limits of these concepts, much elegant theoretical and experimental work has been devoted to finding particularly long or particularly short bonds.[35]

If the HOMO of a molecule is large localized in one bond of a molecule, removal of an electron will lead to substantial lengthening of this bond in the radical cation. This is particularly true for single bonds, where the effect will be larger than for the case of double or triple bonds. However, this effect is only rarely observed since double and triple bonds are much easily oxidized than saturated hydrocarbons. One of the cases where saturated hydrocarbon radical cations can be readily studied is bicyclo[1.1.0]butane **20**, shown in Fig. 8. Here, the HOMO is localized in the central carbon–carbon bond, which lengthens by 0.21 Å upon one-electron oxidation, as shown in Table 1. As pointed out by Bally,[36] state symmetry arguments prevent a complete breakage of the bond. Furthermore, this bond can be varied over a relatively wide range of 0.18 Å by substituent effects. As shown in Table 1, geminal substitution by the sterically demanding ^{t}butyl substituents in **21** and **22** leads to a small contraction of the central carbon–carbon bond relative to the unsubstituted case due to a Thorpe–Ingold effect. Substitution at the bridgehead carbons in **23** leads to a substantial lengthening of this bond by 0.06 Å due to steric repulsion and better stabilization of the localized spin and charge at the bridgehead carbons through hyperconjugation. Finally, the most highly substituted case **24** shows the

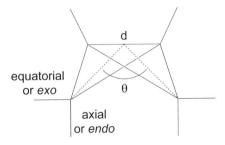

Fig. 8   Geometric parameters in [1.1.0] bicyclobutane **20**.

Table 1 Selected geometrical values for [1.1.0] bicyclobutanes **20–24** and their radical cations (Ref 36b)

		20	21	22	23	24
d (Å)	Neutral	1.49	1.52	1.54	1.48	1.52
	Radical cation	1.70	1.66	1.67	1.76	1.84
θ (°)	Neutral	122.2	133.2	145.8	116.8	123.2
	Radical cation	135.3	141.7	153.3	132.8	143.4

biggest effect with a bond that is elongated by 0.14 Å relative to the unsubstituted case. Unlike the heteroanalog of **24** described by Bertrand,[37] the steric repulsion is, even in this case, not high enough to planarize this system, even though the puckering angle θ increases by more than 20° upon electron transfer to form **24**$^{\cdot+}$. This emphasizes again the importance of state symmetry, which disfavors the open, planar form of the radical cation.

It is clear from these results that the length of a carbon–carbon single bond in a radical cation can be very different if the HOMO has a significant contribution in this bond. For the example discussed here, these bonds can be elongated by as much as 0.32 Å upon ET. Other examples include reactive species such as the acyclic intermediate in radical cation Diels–Alder reaction, which have significantly elongated carbon–carbon single bonds as compared to their neutral, biradical counterparts.[42] At the same time, these partial bonds can be varied over a wider range by substituent effects.[17]

6   Reaction mechanisms

The elucidation of reaction mechanisms is a central topic in organic chemistry that led to many elegant studies emphasizing the interplay of theory and experiment as demonstrated, for example, by the seminal contributions of the Houk group to the understanding of the Diels–Alder and other pericyclic reactions.[38] This reaction class is rather typical for the elucidation of reaction mechanisms. On the experimental side, the toolbox of solvent, substituent and isotope effect studies as well as stereochemical probes have been used extensively, while the reactants, products, intermediates and transition structures involved have been calculated at all feasible levels of theory. As a result, these reactions often serve as a success story in physical organic chemistry.

In comparison, the level of detail in the understanding of radical ion reaction mechanisms is much lower for a number of reasons. Due to the inherently complex nature of the electron transfer–chemical reaction–electron transfer (ECE) mechanism, measurement of substituent, solvent and isotope effects will usually provide a combination of effects on all the steps involved. Introducing a donor substituent on a substrate will, for example, not only change the relative stability of the transition structures and intermediates with localized charges, but will also affect the rate constant of electron transfer and self-exchange between two substrates as well as the rate of back electron transfer.

Computational studies of radical cation pericyclic reactions are also much more difficult than their neutral counterparts. Besides the problems of computational accuracy and unusual electronic effects discussed in an earlier chapter, the reaction pathways for a bimolecular reaction such as the Diels–Alder reaction will be much more complex than in their neutral counterparts. Since the quality of a computational study of a reaction mechanism relies on comparing the relative energies of the relevant pathways computed as unbiased as possible, special care needs to be taken to find and characterize all possible stationary points for a given reaction. There will be, for example, multiple pathways leading to the same products with activation energies that are energetically close enough that the preferred pathway and indeed the overall shape of the reaction hypersurface will be a function of the computational method used. Furthermore, the close energetic spacing of different transition structures can lead to changes in the reaction mechanism as a result of small changes in the substrate. Thus, the notion of a single reaction pathway might not be useful when considering hydrocarbon radical cation reactivity.

An example for multiple, energetically close reaction pathways is the radical cation Diels–Alder reaction. Due to the synthetic importance of the neutral and radical cation equivalent of the Diels–Alder reaction, it is one of the few bimolecular radical cation reactions that have been studied in some detail using both experimental and computational methods. Using spectroscopic techniques[39] and stereochemical probes,[40] a stepwise mechanism was found for a number of radical cation Diels–Alder reactions. Building on earlier studies using UHF and MNDO/3 calculations, the parent reaction of the Diels–Alder reaction of the 1,3-butadiene radical cation with ethylene was investigated using MP3 calculations that indicated a concerted, highly asynchronous reaction.[41] More recently, the same reaction was studied independently by two groups at the QCISD(T)//QCISD and CCSD(T)/MP2 as well as the B3LYP level of theory using double-zeta type basis sets.[42] Fig. 9 summarizes the findings of these studies.

Similar to the pathways discussed earlier, the symmetric concerted pathway involving $26^{•+}$ is subject to a Jahn–Teller distortion and could be only located

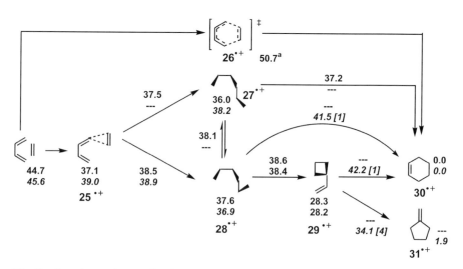

Fig. 9 Reaction pathways for the cycloaddition of the 1,3-butadiene radical cation with ethane. Results from QCISD(T)/6-31G*//QCISD/6-31G* calculations in plain text,[42a] results from UCCSD(T)/DZP//UMP2/DZP in italics.[42b,c] Numbers in brackets indicate the number of additional minima in the pathway.

using the B3LYP/6-31G* level of theory. The activation energy for this process is with 19 kcal/mol at this level quite high and $26^{·+}$ is thus not expected to be part of the reaction pathway. Instead, the butadiene radical cation and ethane form ion–molecule complex $25^{·+}$ with a binding energy of ~7 kcal/mol in the gas phase, which is highly fluctuational in structure. This ion–molecule complex can collapse to one of two singly linked intermediates $27^{·+}$ and $28^{·+}$ in which the ethylene unit is *anti* or *gauche* to the butadiene moiety, respectively. The formation of the singly linked intermediates is only weakly exothermic and occurs with barriers of less than 1.5 kcal/mol. At the same time, the interconversion of $27^{·+}$ and $28^{·+}$ is expected to be very fast. Complex $25^{·+}$, and intermediates $27^{·+}$ and $28^{·+}$ can therefore be thought of as being points on a plateau area of the potential energy hypersurface which can rapidly interconvert and have exit channels to different products. The relative pathways of these three structures leading to the different possible products are highly dependent on the computational method used. The lowest energy pathway calculated at the QCISD(T)/QCISD level of theory is the highly exothermic collapse of $27^{·+}$ to give the cyclohexene radical cation $30^{·+}$ with an activation energy of 1.2 kcal/mol. The related direct addition pathway is predicted to be barrierless based on single point calculations of a UHF/6-31G* IRC calculation.[41] In comparison, $28^{·+}$ can serve as a starting point of a variety of complex reaction pathways. Ring closure to form the vinylcyclobutane radical cation $29^{·+}$ proceeds with an activation energy of only 1 kcal/mol and a reaction energy of -9.7 kcal/mol. The formation of vinylcyclobutane has been observed in a number of reactions of diene radical cations with olefins. ET catalyzed 1,3-sigmatropic shifts can then convert the vinylcyclobutane radical cations into the cyclohexene radical cation $30^{·+}$, resulting in the so-called "indirect radical cation Diels–Alder reaction".[43] Calculations at the UCCSD(T)/DZP//UMP2/DZP level of theory by Hofmann *et al.* indicated a stepwise pathway with an activation energy of 14 kcal/mol for this conversion, while a relatively complex pathway leading to the methylene cyclopentane radical cation $31^{·+}$ has an activation energy of only 5.9 kcal/mol. While products derived from $31^{·+}$ have to the best of our knowledge never been observed in solution phase reactions, mass spectroscopic and computational studies on the cycloaddition of $2^{·+}$ and acetylene provided evidence for a number of different fulvene structures.[44] It is indicative of many hydrocarbon radical cation reactions that the potential energy hypersurfaces obtained in that study intersect with the ones for the $C_6H_8^{·+}$ reactions discussed in Section 2. The finding that several transition structures and intermediates on the hypersurface are so close in energy that different levels of theory will give qualitatively different representations of the hypersurface is also very common. In such cases, no conclusive, quantitatively accurate statement about the reaction pathway can be made. Rather, it is more likely that many different pathways are possible, and that the partitioning between these pathways is controlled by dynamical factors. It is therefore more useful to consider a series of closely related pathways on a plateau rather than a single, well-defined pathway.

One example for such behavior is again the $C_4H_6^{·+}$ hypersurface. Several reactions compete on this hypersurface, some of which have already been discussed in previous chapters. The relative energies of the transition structures and intermediates on this hypersurface are very similar. Bally and coworkers[18] described the different reaction pathways as parts of the so-called "Bauld plateau", indicating the geometric relationship of the different pathways on the plateau with the cyclopropyl carbinyl radical cation **4**$^{\cdot+}$ first described as an intermediate for the ring opening of **1**$^{\cdot+}$ by Bauld.[16] As shown in Fig. 10, the Bauld plateau can be defined by two geometric parameters: a bond distance r_{13} and a dihedral angle θ. This representation clearly demonstrates the strong dependence of the computationally predicted pathway from the method chosen. While QCISD and B3LYP calculations give very similar results for the pathway from **20**$^{\cdot+}$ to **2**$^{\cdot+}$, the B3LYP optimized geometry of the transition structure leading from **1**$^{\cdot+}$ to *trans*-**2**$^{\cdot+}$ is much looser than the geometry calculated at the QCISD level of theory. UMP2 calculations give, however, a qualitatively different picture. Two different pathways leading from **1**$^{\cdot+}$ to *cis*-**2**$^{\cdot+}$ are located: one concerted pathway and one stepwise pathway involving the cyclopropyl carbinyl radical cation **4**$^{\cdot+}$. In addition, an additional pathway connecting **1**$^{\cdot+}$ to **20**$^{\cdot+}$ was found. The finding that all pathways are within a few kcal/mol of each other was attributed to the weaker bonding in

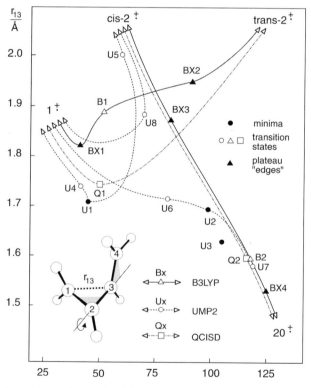

Fig. 10   Schematic sketch of the Bauld plateau (adapted from Ref. 18).

the radical cations as well as low-lying excited state, which interact with the ground state through vibronic interactions. The resulting splitting then maximizes the gap between the surfaces and flattens the ground state surface.

The elucidation of the reaction mechanism of a radical cation reaction is by no means trivial since experimental studies necessarily yield convoluted data and computational studies will for most chemically relevant systems not be accurate enough to distinguish between energetically close pathways. However, the combination of both approaches can provide detailed insights into the mechanisms of the reactions of radical cations. One example where this approach has been successful is the electron transfer catalyzed cycloaddition of indole **32** to 1,3-cyclohexadiene **6**, shown in Fig. 11.[45] This reaction yields after acylation of the initial Diels–Alder adduct **35** *endo* and *exo-***36** in a 3:1 ratio. Using a stereochemical probe, this reaction was shown to be stepwise.[40a] Although the reaction can proceed either through attack of **6** at the 3-position of the indole radical cation **32**$^{\cdot+}$, leading to intermediate **33**$^{\cdot+}$, or initial attack at the 2-position, leading to intermediate **34**$^{\cdot+}$, qualitative considerations as well as low-level calculations[46] indicated that the former pathway is preferred.

These findings were confirmed by calculations at the B3LYP/6-31G* level of theory[47] which favor the pathway involving **33**$^{\cdot+}$ by approximately 3–4 kcal/mol. As can be seen from the results for this pathway, summarized in Fig. 12, the reaction proceeds by initial formation of an ion–molecule complex **37**$^{\cdot+}$ which then leads to the formation of the intermediates *endo* and *exo-***33**$^{\cdot+}$ through transition structures *endo* and *exo-***38**$^{\cdot+}$. The bond lengths in the stereoisomeric transition structures and the singly linked intermediates are very similar. The products *endo* and *exo-***35**$^{\cdot+}$ are then formed through transition structures *endo* and *exo-***39**$^{\cdot+}$. The computed energy difference between the *endo* and *exo* pathway of 0.8 kcal/mol quantitatively reproduces the experimentally observed *endo/exo* ratio of 3:1. Interestingly, the calculation predict different rate determining steps for the *endo* and *exo* pathway. Although this can, in analogy to the smaller reaction energy of the *exo* pathway, be rationalized through the larger steric repulsion in *exo-***39**$^{\cdot+}$, the accuracy of the calculation is not high enough to make a definitive statement regarding the rate determining step. Furthermore, no information on the partitioning ratio between electron transfer and chemical steps can be extracted from these calculations.

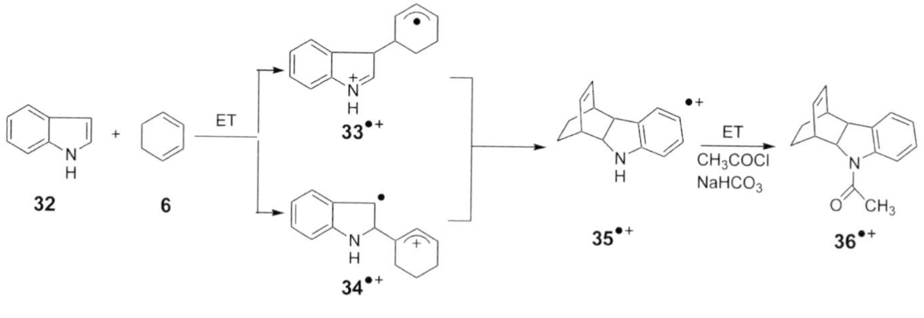

Fig. 11 ETC cycloaddition of indole **32** to 1,3-cyclohexadiene **6**.

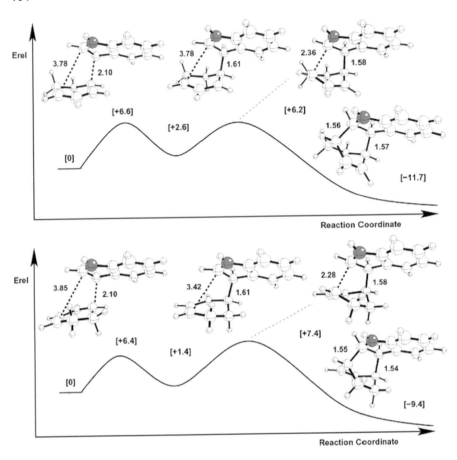

Fig. 12 B3LYP/6-31G* reaction pathways for the *endo* (top) and *exo* (bottom) cycloadditions of Indole **32**$^{\cdot+}$ to 1,3-cyclohexadiene **6**.

The heavy atom isotope effects for this reaction were determined at natural abundance by NMR methodology to be 1.001–1.004 and 1.013–1.016 at C_2 and C_3 of the indole **32**, respectively.[48] The isotope effects calculated from the Biegeleisen–Mayer equation[49] using the frequencies from the B3LYP/6-31G* frequency analysis under the assumption of a 3:1 *endo/exo* partitioning are with 1.005 and 1.026, respectively, which is qualitatively, but not quantitatively correct and indicate a rate-limiting attack at C_3. Quantitative agreement with experiment can be achieved by considering a partitioning between the cycloaddition step from the ion–molecule complex **37**$^{\cdot+}$ to the product **35**$^{\cdot+}$, which proceeds with a rate constant k_1 and an electron exchange between **32** and **37**$^{\cdot+}$ with a rate constant k_2.

$$\mathbf{37}^{\cdot+} \xrightarrow{k_1} \mathbf{35}^{\cdot+} \tag{1}$$

$$37^{\cdot+}_{12C} + 32^{\cdot+}_{13C} \overset{k_2}{\rightarrow} 37^{\cdot+}_{13C} + 32^{\cdot+}_{12C} \qquad (2)$$

Numerical simulations of the isotope effects involved indicate that it is unlikely that k_2 is fast enough to make k_1 rate limiting. In order to predict the overall isotope effect of the reaction, reasonable assumptions on the values of k_1 and k_2 need to be made. The results of the numerical simulation can be then compared to the experimental results. Using $k_5 = 10^8 \ M^{-1} \ s^{-1}$ and $k_3 = 10^7 \ s^{-1}$, the predicted isotope effects for C_2 and C_3 are reduced to 1.003 and 1.014, respectively. Although this agreement with the experimental results does not provide definitive values for the rates of electron exchange k_2 and cycloaddition k_1, it does show that both steps need to be considered to achieve consistency between the experimental and calculated results. In particular, it is unlikely that k_2 will become large enough to make the chemical step with the rate constant k_1 exclusively rate determining. Furthermore, it demonstrates that despite the difficulties in elucidating radical ion mechanisms, the combination of experimental and computational methods can provide detailed insights.

7 Conclusions

The examples discussed in this review demonstrate that although the chemistry of hydrocarbon radical cations can be understood in terms of the common concepts of physical organic chemistry, the relative importance of these concepts can be quite different from what is often expected. The interplay of symmetry, changes in bonding characteristics and the need to stabilize the highly reactive radical cation intermediates through conjugation can lead to intriguing and often complex reaction pathways that are not easily anticipated based on the knowledge of the mechanism of their neutral counterparts. Therefore, quantitative electronic structure methods in combination with modern experimental tools are useful for the mechanistic investigations of such reactions. The differences in structure and reactivity between neutral, closed-shell organic compounds and their radical cation counterparts highlighted here as well as the generally low activation energies also indicate the potential for developing new synthetic methodology or to uncover biochemical reaction mechanisms that use electron transfer catalysis.

Acknowledgements

We gratefully acknowledge the financial support of our work by the National Institutes of Health (CA73775), the National Science Foundation (CHE-9733050), and the Volkswagen Foundation (I/72 647) as well as a Camille Dreyfus Teacher–Scholar Award to O.W. Our own work benefited greatly from discussions with T. Bally (Fribourg), H. Hopf (Braunschweig), D. Schroeder (Berlin), and

D. A. Singleton (Texas A&M). Most importantly, we would like to thank K. N. Houk, to whom this review is dedicated, for his contributions to physical organic chemistry without which the work in our group would have been impossible.

References

1. For reviews of classical treatments of radical cation chemistry, compare: (a) Kaiser, T.E. (1968). *Radical Ions*. Wiley-Interscience, New York; (b) Roth, H.D. (1986). *Tetrahedron* **42**(22), 6097–6100; (c) Roth, H.D. (1990). *Top. Curr. Chem.* **156**, 1
2. For some recent examples of electron transfer induced reaction in organic synthesis, compare; (a) Rössler, U., Blechert, S. and Steckhan, E. (1999). *Tetrahedron Lett.* **40**, 7075; (b) Jonas, M., Blechert, S. and Steckhan, E. (2001). *J. Org. Chem.* **66**, 6896; (c) Moeller, K.D. (2000). *Tetrahedron* **56**, 9527; (d) Liu, B. and Moeller, K.D. (2001). *Tetrahedron Lett.* **42**, 7163; (e) Kumar, V.S. and Floreancig, P.E. (2001). *J. Am. Chem. Soc.* **123**, 3842; (f) Kumar, V.S., Aubele, D.L. and Floreancig, P.E. (2001). *Org. Lett.* **3**, 4123; (g) Duan, S.Q. and Moeller, K.D. (2001). *Org. Lett.* **3**, 2685; (h) Pandey, G. and Kapur, M. (2001). *Synthesis* 1263; (i) Pandey, G., Laha, J.K. and Lakshmaiah, G. (2002). *Tetrahedon* **58**, 3525; (j) Goeller, F., Heinemann, C. and Demuth, M. (2001). *Synthesis* 1114; (k) Cocquet, G., Rool, P. and Ferroud, C. (2001). *Tetrahedron Lett.* **42**, 839; (l) Bertrand, S., Hoffman, N. and Pete, J.-P. (2000). *Eur. J. Org. Chem.* 2227; (m) Bertrand, S., Glapski, C., Hoffman, N. and Pete, J.-P. (1999). *Tetrahedron* **52**, 3425
3. Fokin, A.A. and Schreiner, P.R. (2002). *Chem. Rev.* **102**, 1551
4. (a) Schenck, C.C., Diner, B., Mathis, P. and Satoh, K. (1982). *Biochim. Biophys. Acta.* **680**, 216; (b) Mathis, P. and Rutherford, A.W. (1984). *Biochim. Biophys. Acta.* **767**, 217; (c) Moore, T.A., Gust, D., Mathis, P., Mialocq, J.-C., Chachaty, C., Benassom, R.V., Land, E.J., Doizi, D., Lidell, P.A., Lehman, W.R., Nemeth, G.A. and Moore, A.L. (1984). *Nature* **307**, 630; (d) Gust, D., Moore, T.A., Lidell, P.A., Nemeth, G.A., Moore, A.L., Barrett, D., Pessiki, P.J., Benassom, R.V., Rougee, M., Chachaty, C., DeSchryver, F.C., van der Auweraer, M., Holzwarth, A.R. and Conolly, J.S. (1987). *J. Am. Chem. Soc.* **109**, 846; (e) Boll, M., Laempe, D., Eisenreich, W., Bacher, A., Mittelberger, T., Heinze, J. and Fuchs, G. (2000). *J. Biol. Chem.* **275**, 21889; (f) Unciuleac, M. and Boll, M. (2001). *Proc. Natl. Acad. Sci. USA.* **98**, 13619; (g) Buckel, W. and Golding, B.T. (1999). *FEMS Microbiology Reviews* **22**, 523; (h) Hans, M., Bill, E., Cirpus, I., Pierik, A.J., Hetzel, M., Alber, D. and Buckel, W. (2002). *Biochemistry* **41**, 5873
5. (a) Bauld, N.L. (1989). *Tetrahedron* **45**, 5307; (b) Chanon, M. and Eberson, L. (1989). *Photoinduced Electron Transfer Part A*. In, Chanon, M. and Fox, M.A. (eds), pp. 409. Elsevier, Amsterdam; (c) Eberson, L. (1987). *Electron Transfer Reactions in Organic Chemistry*, Springer, Berlin; (d) Bauld, N.L. (1992). In *Advances in Electron Transfer Chemistry*. Mariano, P.S. (ed.), vol. 2, pp. 1. Jai Press, New York; (e) Schmittel, M. and Burghart, A. (1997). Angew. Chem. Int. Ed. Eng. 36, 2551; (f) Pandey, G. (1993). Top. Curr. Chem. 168, 175; (g) Moeller, K.D., Liu, B., Reddy, S.H.K., Sun, H., Sun, Y., Sutterer, A. and Chiba, K. (2001). Proc. Electrochem. Soc. 2001; (h) Cossy, J. and Pete, J.-P. (1996). In Advances in Electron Transfer Chemistry, Mariano, P.S. (ed.), vol. 141. JAI Press, New York; (i) Saettel, N.J., Oxgaard, J. and Wiest, O. (2001). Europ. J. Org. Chem. 1429
6. For recent books on the application of ETC and radical ions in organic chemistry, compare: (a) Mattay, J. and Astruc, D. (eds), (2001). *Organic and Metalorganic Systems. Electron Transfer in Chemistry*. vol. 2. Wiley-VCH, New York; (b) Linker, T. and Schmittel, M. (1998). *Radikale und Radikalionen in der Organischen Synthese*. Wiley-VCH, New York; (c) Bauld, N.L. (1997). *Radicals, Ion Radicals, and Triplets*. Wiley-VCH, New York

7. For a more comprehensive discussion of the different approaches to the calculation of radical cations, compare: Bally, T. and Borden, W.T. (1999). In *Reviews in Computational Chemistry*, Lipkowitz, K. and Boyd, D. (eds), vol. 13, pp. 1–71. Wiley/VCH, New York

8. See for example: (a) Hrouda, V., Čársky, P., Ingr, M., Chval, Z., Sastry, G.N. and Bally, T. (1998). *J. Phys. Chem. A* **102**, 9297; (b) Hrouda, V., Roeselova, M. and Bally, T. (1997). *J. Phys. Chem. A* **101**, 3925; (c) Wiest, O. (1996). *J. Mol. Struct (THEOCHEM)* **368**, 39; (d) Ma, N.L., Smith, B.J. and Radom, L. (1992). *Chem. Phys. Lett.* **193**, 386; (e) Nobes, R.H., Moncrieff, D., Wong, M.W., Radom, L., Gill, P.M.W. and Pople, J.A. (1991). *Chem. Phys. Lett.* **182**, 216; (f) Mayer, P.M., Parkinson, C.J., Smith, D.M. and Radom, L. (1998). *J. Chem. Phys.* **108**, 604

9. (a) Noodleman, L., Post, D. and Baerends, E. (1982). *J. Chem. Phys.* **64**, 159; (b) Sodupe, M., Bertran, J., Rodriguez-Santiago, L. and Baerends, E.J. (1999). *J. Phys. Chem. A* **103**, 166; (c) Braida, B., Hiberty, P.C. and Savin, A. (1998). *J. Phys. Chem. A*. **102**, 7872

10. Bally, T. and Sastry, G.N. (1997). *J. Phys. Chem. A*. **101**, 7923

11. Jahn, H.A. and Teller, E. (1937). *Proc. R. Soc.* **161**, 220

12. For a discussion of Jahn–Teller effects, compare: Bersuker, I.B. (2001). *Chem. Rev.* **101**, 1067

13. Wiest, O. (1997). *J. Am. Chem. Soc.* **119**, 5713

14. Aebischer, J.N., Bally, T., Roth, K., Haselbach, E., Gerson, F. and Qin, X.-Z. (1989). *J. Am. Chem. Soc.* **111**, 7909

15. For other experimental studies of the ring opening of substituted cyclobutene radical cations: (a) Gross, M.L. and Russell, D.H. (1979). *J. Am. Chem Soc.* **101**, 2082; (b) Dass, C., Sack, T.M. and Gross, M.L. (1984). *J. Am. Chem. Soc.* **106**, 5780; (c) Dass, C. and Gross, M.L. (1983). *J. Am. Chem. Soc.* **105**, 5724; (d) Haselbach, E., Bally, T., Gschwind, R., Hemm, U. and Lanyiova, Z. (1979). *Chimia* **33**, 405; (e) Kawamura, Y., Thurnauer, M. and Schuster, G.B. (1986). *Tetrahedron* **42**, 6195; (f) Brauer, B.-E. and Thurnauer, M.C. (1987). *Chem. Phys. Lett.* **113**, 207; (g) Gerson, F., Qin, X.Z., Bally, T. and Aebischer, J.N. (1988). *Helv. Chim. Acta.* **71**, 1069; (h) Miyashi, T., Wakamatsu, K., Akiya, T., Kikuchi, K. and Mukai, T. (1987). *J. Am. Chem. Soc.* **109**, 5270; (i) Takahaski, Y. and Kochi, J.K. (1988). *Chem. Ber.* **121**, 253

16. Bellville, D.J., Chelsky, R. and Bauld, N.L. (1982). *J. Comp. Chem.* **3**, 548

17. Swinarski, D.J. and Wiest, O. (2000). *J. Org. Chem.* **65**, 6708

18. Sastry, G.N., Bally, T., Hrouda, V. and Čársky, P. (1998). *J. Am. Chem. Soc.* **120**, 9323

19. Barone, V., Rega, N., Bally, T. and Sastry, G.N. (1999). *J. Phys. Chem. A* **103**, 217

20. (a) Cave, R. and Johnson, J.L. (1992). *J. Phys. Chem.* **96**, 5332; (b) Kesztheli, T., Wilbrandt, R., Cave, R. and Johnson, J.L. (1994). *J. Phys. Chem.* **98**, 6532; (c) Fülscher, M.P., Matzinger, S. and Bally, T. (1995). *Chem. Phys. Lett.* **236**, 167; (d) Kawashima, Y., Nakayama, K., Nakano, H. and Hirao, K. (1997). *Chem. Phys. Lett.* **267**, 82; (e) Bally, T., Nitsche, S., Roth, K. and Haselbach, E. (1984). *J. Am. Chem. Soc.* **106**, 3927

21. Radosevich, A.T. and Wiest, O. (2001). *J. Org. Chem.* **66**, 5808

22. However, the ring closure of substituted hexatriene radical cations has been reported: (a) Barkow, A. and Grützmacher, H.-F. (1994). *Intl, J. Mass Spectrom Ion Proc.* **142**, 195; It should also be noted that the ring closure of a hexatriene radical anion has also been reported: (b) Fox, M.A. and Hurst, J.R. (1984). *J. Am. Chem. Soc.* **106**, 7626

23. (a) Kelsall, B.J. and Andrews, L. (1984). *J. Phys. Chem.* **88**, 2723; (b) Bally, T., Nitsche, S., Roth, K. and Haselbach, E. (1985). *J. Phys. Chem.* **89**, 2528; (c) Shida, T., Kato, T. and Nosaka, Y. (1977). *J. Phys. Chem.* **81**, 1095

24. See for example; Wheland, G.W. (1955). *Resonance in Organic Chemistry*. Wiley, New York

25. See for example: (a) Keszthelyi, T. and Wilbrandt, R.T. (1997). *J. Mol. Struct.* **410**, 339; (b) Bally, T., Roth, K., Tang, W., Schrock, R.R., Knoll, K. and Park, L.Y. (1992). *J. Am. Chem. Soc.* **114**, 2440; (c) Shirakawa, H. (2001). *Angew. Chem. Int. Ed. Eng.* **40**, 2574

26. Foresman, J.B., Wong, M.W., Wiberg, K.B. and Frisch, M.J. (1993). *J. Am. Chem. Soc.* **115**, 2220

27. Oxgaard, J. and Wiest, O. (2001). *J. Phys. Chem. A* **105**, 8236

28. (a) Mayr, H., Forner, W. and Schleyer, P.v.R. (1979). *J. Am. Chem. Soc.* **101**, 6032; (b) Rahavachari, K., Whiteside, R.A., Pople, J.A. and Schleyer, P.v.R. (1981). *J. Am. Chem. Soc.* **103**, 5649; (c) Cournoyer, M.E. and Jorgensen, W.L. (1984). *J. Am. Chem. Soc.* **106**, 5104

29. (a) Gobbi, A. and Frenking, G. (1994). *J. Am. Chem. Soc.* **116**, 9275; (b) Mo, Y., Lin, Z., Wu, W. and Zhang, Q. (1996). *J. Phys. Chem.* **100**, 6469

30. (a) Wiberg, K.B., Breneman, C.M. and LePage, T.J. (1990). *J. Am. Chem. Soc.* **112**, 61; (b) Mo, Y. and Peyerimhoff, S.D. (1998). *J. Chem. Phys.* **109**, 1687

31. (a) Feller, D., Davidson, E.R. and Borden, W.T. (1984). *J. Am. Chem. Soc.* **106**, 2513; (b) Karadakov, P.B., Gerratt, J., Raos, G., Cooper, D.L. and Raimondi, M. (1994). *J. Am. Chem. Soc.* **116**, 2075

32. (a) Tsuzuki, S., Schaefer, L., Hitoshi, G., Jemmis, E.D., Hosoya, H., Siam, K., Tanabe, K. and Osawa, E. (1991). *J. Am. Chem. Soc.* **113**, 4665; (b) Karpfen, A. (1999). *J. Phys. Chem.* **103**, 2821

33. (a) Traetteberg, M., Hopf, H., Lipka, H. and Hanel, R. (1994). *Chem. Ber.* **127**, 1459; (b) Traetteberg, M., Bakken, P., Hopf, H. and Hanel, R. (1994). *Chem. Ber.* **127**, 1469

34. Oxgaard, J. and Wiest, O. (2002). *J. Phys. Chem. A.* **106**, 3967

35. For an interesting overview, see: Hopf, H. (2000). *Classics in Hydrocarbon Chemistry.* VCH-Wiley, Weinheim

36. Bally, T. (1991). *J. Mol. Struct (THEOCHEM)* **227**, 249; Compare also (b) Saettel, N. J. and Wiest, O. (2003). *J. Org. Chem.* 68 ASAP

37. Scheschkewitz, D., Amii, H., Gornitzka, H., Schoeller, W.W., Bourissou, D. and Bertrand, G. (2002). *Science* **295**, 1880

38. For overviews of these contributions, compare e.g.: (a) Dolbier, W.R., Jr, Koroniak, H., Houk, K.N. and Sheu, C. (1996). *Acc. Chem. Res.* **29**, 471; (b) Houk, K.N., Gonzalez, J. and Li, Y. (1995). *Acc. Chem. Res.* **28**, 81; (c) Houk, K.N. (1989). *Pure Appl. Chem.* **61**, 643; (d) Borden, W.T., Loncharich, R.J. and Houk, K.N. (1988). *Ann. Rev. Phys. Chem.* **39**, 213; (e) Houk, K.N., Paddon-Row, M.N., Rondan, N.G., Wu, Y.-D., Brown, F.K., Spellmeyer, D.C., Metz, J.T., Li, Y. and Loncharich, R. (1986). *J. Science* **231**, 1108; (f) Houk, K.N. (1983). *Pure Appl. Chem.* **55**, 277; (g) Houk, K.N. (1979). *Top. Curr. Chem.* **79**, 1; (h) Wiest, O. and Houk, K.N. (1996). *Top. Curr. Chem.* **183**, 1; (i) Wiest, O., Montiel, D.C. and Houk, K.N. (1997). *J. Phys. Chem. A.* **101**, 8378

39. (a) Roth, H.D., Schilling, M.L.M. and Abelt, C.L. (1986). *Tetrahedron* **42**, 6157; (b) Roth, H.D. and Schilling, M.L. (1985). *J. Am. Chem. Soc.* **107**, 716; (c) Turecek, F. and Hanus, V. (1984). *Mass. Spectrom Rev.* **3**, 85

40. (a) Wiest, O. and Steckhan, E. (1993). *Tetrahedron Lett.* **34**, 6391; (b) Gao, D. and Bauld, N.L. (2000). *J. Org. Chem.* **65**, 6276; (c) Gao, D. and Bauld, N.L. (2000). *J. Chem. Soc. Perkin Trans.* 931

41. (a) Bellville, D.J. and Bauld, N.L. (1986). *Tetrahedron* **42**, 6167; (b) Bauld, N.L., Bellville, D.J., Pabon, R.A., Chelsky, R. and Green, G.J. (1983). *J. Am. Chem. Soc.* **105**, 2378; (c) Bellville, D.J., Bauld, N.L., Pabon, R.A. and Gardner, S.A. (1983). *J. Am. Chem. Soc.* **105**, 3584; (d) Bauld, N.L. (1992). *J. Am. Chem. Soc.* **114**, 5800

42. (a) Haberl, U., Wiest, O. and Steckhan, E. (1999). *J. Am. Chem. Soc.* **121**, 6730; (b) Hofmann, M. and Schaefer, H.F. (1999). *J. Am. Chem. Soc.* **121**, 6719; (c) Hofmann, M. and Schaefer, H.F. (2000). *J. Phys. Chem. A.* **103**, 8895

43. (a) Pabon, R.A., Bellville, D.A. and Bauld, N.L. (1984). *J. Am. Chem. Soc.* **106**, 2730;
    (b) Reynolds, D.W., Harirchian, B., Chiou, H., Marsh, B.K. and Bauld, N.L. (1989).
    J. Phys. Org. Chem. **2**, 57; (c) Botzem, J., Haberl, U., Steckhan, E. and Blechert, S.
    (1998). *Acta. Chem. Scand.* **52**, 175; (d) Pabon, R.A., Belville, D.J. and Bauld, N.L.
    (1984). *J. Am. Chem. Soc.* **106**, 2730; (e) Bauld, N.L., Harirchian, B., Reynolds, D.W. and
    Whitem, J.C. (1988). *J. Am. Chem. Soc.* **110**, 8111
44. Bouchoux, G., Nguyen, M.T. and Salpin, J.-Y. (2000). *J. Phys. Chem. A.* **104**, 5778
45. (a) Gieseler, A., Steckhan, E. and Wiest, O. (1990). *Synlett* 275; (b) Gieseler, A.,
    Steckhan, E., Wiest, O. and Knoch, F. (1991). *J. Org. Chem.* **56**, 1405
46. (a) Wiest, O., Steckhan, E. and Grein, F. (1992). *J. Org. Chem.* **57**, 4034; (b) Haberl, U.,
    Steckhan, E., Blechert, S. and Wiest, O. (1999). *Chem. Eur. J.* **5**, 2859
47. Saettel, N.J., Wiest, O., Singleton, D.A. and Meyer, M.P. (2002). *J. Am. Chem. Soc.* **124**,
    11552
48. Singleton, D.A. and Thomas, A.A. (1995). *J. Am. Chem. Soc.* **117**, 9357
49. (a) Bigeleisen, J. and Mayer, M.G. (1947). *J. Chem. Phys.* **15**, 261; The isotope effects
    were calculated using QUIVER with a scaling factor of 0.9614 and a correction for
    hydrogen tunneling; (b) Saunders, M., Laidig, K.E. and Wolfsberg, M. (1989). *J. Am.
    Chem. Soc.* **111**, 8989; (c) Scott, A.P. and Radom, L. (1996). *J. Phys. Chem.* **100**, 16502;
    (d) Bell, R.P. (1980). *The Tunnel Effect in Chemistry*, p. 60. Chapman & Hall, London

Charge distribution and charge separation in radical rearrangement reactions

H. Zipse

Department Chemie, LMU München, Butenandstr. 13, D-81377 München, Germany

1   Introduction

Concerted rearrangement or stepwise heterolytic dissociation/recombination? This question has been at the heart of many mechanistic studies involving aliphatic radicals carrying electronegative substituents X adjacent to the radical center (Scheme 1). Even though the debate around the mechanistic options in these open shell systems has never been as dramatic as the one on stepwise or concerted pericyclic reactions,[1,2] a shift in the major mechanistic paradigm has nevertheless occurred in recent years from mainly concerted to mainly stepwise even in apolar solution.[3] The situation is complicated through the presence of additional reaction pathways to those shown in Scheme 1, the dissociation of the contact radical ion pair (CRIP) into free (radical) ions being the most important as it provides the experimental basis for direct detection of the heterolytic process.[3] Indirect proof for the heterolytic character of a particular reaction stems from kinetic analyses or the observation of solvent and substituent effects. Early theoretical studies by Radom *et al.* for migrations of acyloxy groups (X = O(CO)H, Scheme 1) indicate, however, that even a concerted process might be characterized by a substantial degree of charge separation.[4] This implies that the observation of solvent or substituent effects alone is insufficient proof for a stepwise, heterolytic process.

A comprehensive review of experimental as well as theoretical studies has been compiled by Beckwith *et al.* in 1997 for those systems with acyloxy or phosphatoxy substituents.[3] This account will therefore concentrate on the theoretical studies published since then together with the relevant experimental results. Particular

111

ADVANCES IN PHYSICAL ORGANIC CHEMISTRY
VOLUME 38 ISSN 0065-3160   DOI 10.1016/S0065-3160(03)38003-7

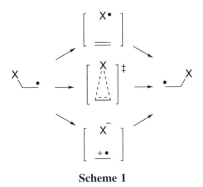

Scheme 1

attention will be paid to quantitative aspects of the charge distribution in ground and transition states, as the fully heterolytic pathway must be expected to develop substantial negative charge on the migrating group X (Scheme 1). In case the C–X bond cleavage process were to proceed in a completely homolytic manner, the migrating group X would accumulate most of the unpaired spin density at some point along the reaction pathway. The (negative) charge of the migrating group X as well as its unpaired spin density will therefore be used in the following to differentiate between the homo- and the heterolytic character of a particular rearrangement reaction. Even though not reflecting the geometrical aspects of charge separation, the term heterolytic will be used to indicate pathways with enhanced negative charge on the migrating group X, while the term homolytic will be used for transition states with enhanced unpaired spin density at X.

2 β-Haloalkyl radicals

The simple most systems displaying the structural motive described in Scheme 1 carry a halogen atom in β-position to the radical center. An early controversy arising from stereochemical experiments deals with the equilibrium structure of these types of radicals.[5,6] The stereochemical control observed in some of these reactions suggests that the halogen is either asymmetrically or symmetrically bridging the radical center, in particular if X = Br or I (Scheme 2).

While recent theoretical studies indeed support the hypothesis of a symmetrically bridged intermediate for X = Br and I, the situation is less clear for X = Cl as the

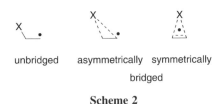

unbridged   asymmetrically   symmetrically
bridged

Scheme 2

existence of a symmetric intermediate depends on the choice of the theoretical method.[6,7] In all theoretical studies, however, the symmetric structure (being a transition state or a true intermediate) is energetically less favorable than the unbridged minimum for $X = Cl$. Table 1 contains an overview of energy differences between the unbridged minimum energy structure and the symmetrically bridged structure, the latter being a transition state in most but not all cases. Positive energy differences indicate a preference for the unbridged structure.

The energy differences compiled in Table 1 clearly illustrate that calculation of the reaction barrier for the 1,2-migration process through a symmetrically bridged intermediate or transition state is by no means a trivial task. The very different results obtained with seemingly similar methods may be due to the fact that several different electronic states exist for the C_{2v} symmetrically bridged structure **2a** ($R = H$). It is interesting to see that the MRDCI results obtained by Engels *et al.* can be reproduced by some of the all-electron hybrid density functional methods. Unfortunately, the charge and spin density distribution has not been characterized at most of these levels. According to the Natural Population Analysis (NPA)[9] scheme the chlorine atomic charge (q) amounts to -0.13 in the unbridged ground state **1a** and to -0.27 in the symmetrically bridged transition state **2a** while the unpaired spin density (SD) values on chlorine for these structures are 0.15 and 0.68, respectively, at the UB3LYP/6-31G(d) level of theory.[8] The small rise in negative partial charge of the chlorine atom on proceeding from the ground to the transition state for the 1,2-migration process as well as the much larger increase in unpaired spin density is indicative of a mixed homo/heterolytic bond cleavage process with a dominating homolytic component.

The chlorine 1,2-migration process has also been studied in a slightly larger model system, the 3-chloro-2-butyl radical **1b** (Scheme 3, $R = CH_3$).[7] Again the barrier for chlorine 1,2-migration is strongly dependent on the level of theory, the contribution of exact exchange in hybrid DFT methods being particularly important. If results for identical theoretical levels are compared it becomes clear that introduction of the two methyl groups lowers the reaction barrier quite significantly. The chlorine partial charge as calculated with the NPA scheme at the UB3LYP/6-31G(d)//UB3LYP/aug-cc-pVDZ level of theory now amounts to -0.34 in **2b** and -0.19 in **1b** while the spin density values at chlorine for these structures are 0.61 and 0.17, respectively. Comparison to the values obtained for **1a/2a** shows only minor changes, despite the substantial variation in reaction barrier.

The effects of electron-withdrawing substituents have been studied by Goddard *et al.* who showed that the chlorine 1,2-migration barrier rises by 42.7 kJ/mol on exchange of all hydrogen atoms of the parent 2-chloroethyl radical **1a** by fluorine atoms.[6i] Even larger effects have been found for the analogous bromine 1,2-migration process.[6i,10] The theoretically predicted preference of the unbridged structure in these latter cases has recently been supported by ultrafast gas phase electron diffraction measurements.[11,12]

The development of the chlorine charge and spin densities along the migration pathway can much better be appreciated in a graphical representation plotting the

Table 1 Energy differences between unbridged radical **1** and symmetrically bridged radical **2** (in kJ/mol)

Level of theory	ΔE_0 (kJ/mol)^a	Reference
R = H		
MRDCI	$+25-27^b$	6c
PMP2/6-31G(d,p)	$+273$	6g
UMP2/DZP + BF	$+55.6^b$	6d
LMP2/LAV3P	$+2.3$	6e
UB3LYP/aug-cc-pVDZ	$+25.6\ (+24.8)^c$	8
UB3LYP/6-31G(d,p)	$+26.4$	6g
UB3LYP/6-31G(d)	$+27.4\ (+26.7)^c$	8
B3LYP/LAV3P	$+7.5$	6e
B3PW91/LAV3P	$+19.3$	6e
R = CH$_3$		
G3(MP2)B3	$+30.2\ (+29.2)^c$	7
UB3LYP/aug-cc-pVDZ	$+11.5\ (+10.5)^c$	7
UBHLYP/aug-cc-pVDZ	$+27.4\ (+26.5)^c$	7

^{a}Relative total energies including zero point vibrational energy differences.
^{b}Relative total energies without zero point energy correction.
^{c}Relative total enthalpies at 298 K.

chlorine charge density values along one axis and the chlorine spin densities along the second (Fig. 1). This type of representation may be termed a spin density/charge density plot (in short: sdq-plot) and is reminiscent of the More O'Ferrall-Jencks bond order diagrams.[13] As in the latter it may be helpful to illustrate the corners of the sdq-plot with limiting valence bond configurations of integral chlorine charge or spin density values. While the situation of zero charge and zero spin density at chlorine can best be characterized with the Lewis structure shown in the lower right of Fig. 1a, a strictly homolytic cleavage of the C–Cl bond would lead along the vertical axis to the Lewis structure in the upper right. Alternatively, the C–Cl bond can be cleaved heterolytically along the horizontal axis leading to the Lewis structure in the lower left corner. The two ground states **1a** and **1b** are located in the vicinity of the lower right corner of Fig. 1a, indicating only small admixtures of the homo- and heterolytic VB configurations to the covalent ground states. The transition states **2a** and **2b**, on the other hand, are located a good distance away from the respective ground states towards the homolytic Lewis structure in the upper right

a: R = H; **b**: R = CH$_3$

Scheme 3

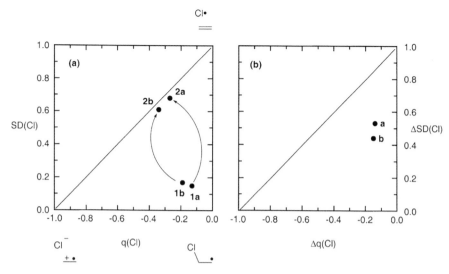

Fig. 1 (a) Graphical representation of the chlorine charge and spin densities for chloralkyl radicals **1a** (R = H) **1b** (R = CH₃) as calculated at the UB3LYP/6-31G(d) level of theory using the NPA scheme. (b) The same data as those in (a) represented as differences in charge and spin densities between ground states **1** and transition states **2**.[14]

corner and only slightly to the left of the ground states in the heterolytic direction. Furthermore, both transition states are located close to the diagonal connecting the heterolytic and the homolytic corners of Fig. 1a, indicating little contribution of the covalent structure in the lower right corner.

How do transition state charge and spin density distribution relate to reaction barriers? One may argue that variations in barrier heights do not correlate with the properties of transition states alone but with the *differences* between ground and transition state properties. A plot of the *differences* between ground and transition state charge and spin densities would therefore be much more appropriate than consideration of the absolute values themselves. To this end a second plot (b) has been included in Fig. 1 indicating the differences in chlorine charge densities Δq and spin densities ΔSD (in short: dsdq-plot).

Differences in charge or spin density can, of course, derive from a variety of different *absolute* values and the dsdq-plot can therefore not be illustrated with limiting Lewis structures as is the case for the alternative sdq-plot. The only point clearly defined in the dsdq-plot in Fig. 1b is the lower right corner which corresponds to the ground state and thus reference point of the system. For the current case of chlorine 1,2-migration in **1a** and **1b** the transition states are displaced relative to the origin of the coordinate system used here towards the homolytic direction much more than to the heterolytic one, reconfirming the characterization as a mainly homolytic process with some heterolytic admixture. The small changes in transition state characteristics together with the large barrier lowering on introduction of the two methyl groups suggests that much larger changes can occur upon introduction of more strongly stabilizing substituents such as aryl groups. For strongly electron-donating substituents it appears possible that transition structure **2** turns into a minimum whose electronic structure resembles that of a CRIP while for strongly electron withdrawing substituents the homolytic character of the bond cleavage process might be enhanced. Tan *et al.* have recently studied a variety of substituted *β*-chloro-benzyl radicals generated in the course of the tin hydride reduction of benzyl bromides and observed a dramatic dependence of the product distribution on the substitution pattern.[15] This *Polar Effects Controlled Enantiose-lective 1,2-Chlorine Atom Migration* is strong support for the mixed homo/heterolytic character of the chlorine migration process and illustrates the possibility of manipulating the participating reaction channels in a semi-rational manner. Similarly, large solvent and substituent effects have been observed in a nanosecond laser flash photolysis study of *β*-halobenzyl radicals by Cozens *et al.*[16] These latter results have been rationalized assuming a competition between homolytic and heterolytic pathways without the involvement of any bridged intermediates.

3   *β*-Acyloxyalkyl radicals

The 1,2-acyloxy migration in *β*-acyloxyalkyl radicals has been the subject of many experimental[3] as well as some theoretical studies.[4,17] While a broad spectrum of mechanisms is conceivable for this kind of process,[3] only three will be discussed here in detail (Scheme 4): (i) stepwise heterolysis/recombination involving

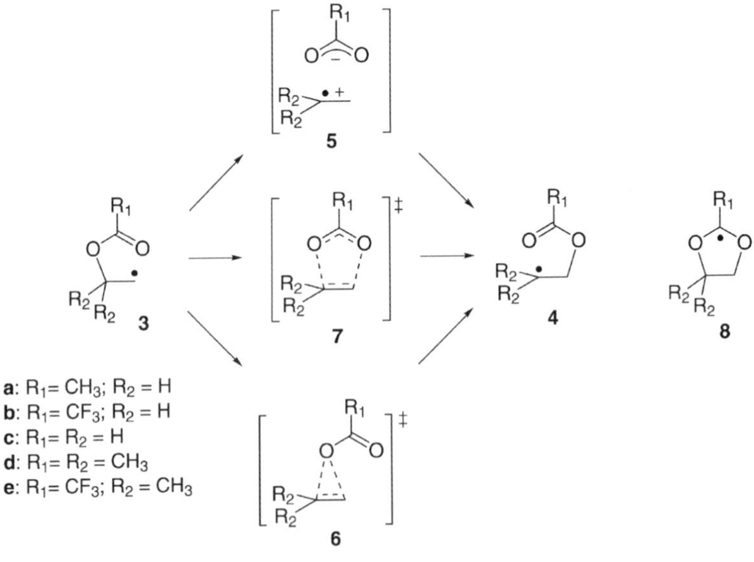

a: $R_1 = CH_3$; $R_2 = H$
b: $R_1 = CF_3$; $R_2 = H$
c: $R_1 = R_2 = H$
d: $R_1 = R_2 = CH_3$
e: $R_1 = CF_3$; $R_2 = CH_3$

Scheme 4

formation of a CRIP intermediate **5**, (ii) concerted [1,2]-migration through a three-membered ring transition state such as **6**, (iii) concerted [3,2]-migration through a five-membered ring transition state such as **7**. A fourth possibility involves cyclization to give the 1,3-dioxolanyl radical **8** and subsequent ring opening to give product radical **4**. However, the intermediacy of dioxolanyl radicals has convincingly been ruled by direct as well as indirect kinetics studies.[18–20] This is in agreement with theoretical studies[4,17] on model systems **3a–3e** described in Scheme 4 and we can thus neglect this stepwise pathway.

Differentiation between the concerted [1,2]- and [3,2]-migration pathways should, in principle, be possible through isotopic labeling experiments at the acyloxy group. Unfortunately, however, it was found that the outcome of these labeling studies depends dramatically on the substitution pattern with faster rearrangement reactions involving more strongly stabilizing substituents having a preference for the [1,2]-process.[20–25] This result could, of course, also result if the CRIP **5** were to play a major role in all of these reactions.

All theoretical studies published to date on model systems **3a–3e** confirm the existence of transition states **6** and **7**, but not that of ion pair intermediates **5**. Table 2 contains an overview of reaction barriers for the two competing pathways as well as charge and spin density data for ground and transition states. The charge and spin densities given here are cumulative values including all partial atomic charges and unpaired spin densities of the migrating acyloxy group. This choice ensures complete comparability with the results obtained for the halogen migration reactions.

The calculation of reaction barriers for the 1,2-migration process again turns out to be a challenge as the B3LYP barriers are consistently lower than those calculated at either QCISD/6-31G(d) or G3(MP2)B3 level, the latter of which may be the most accurate in this comparison. One general trend visible in Table 2 is that substituent effects appear to be larger in the alkyl radical part as compared to the acyl group. If only the B3LYP results are considered, it also appears that the disfavored [1,2]-shift pathway becomes more competitive with lower absolute reaction barriers, that is, in the more highly substituted systems **3d** and **3e**.

Analysis of the cumulative charge and spin densities in Table 2 shows that the migrating acyloxy groups are more negatively charged in ground state **3** as compared to the transition states **6** and **7** in the less reactive systems **3a–3c**. In the more reactive systems with $R_2 = CH_3$, the acyloxy group charge in the transition states is either similar to or even larger than in the ground state. This also implies that variation of the substitution pattern has little influence on the ground state charge distribution, but a much larger effect on the transition states. The graphical representation of the charge and spin density development in acyloxy migrations in Fig. 2 clearly shows that all of the reactions studied here have less heterolytic character than the chlorine migration reactions studied before. In the sdq-plot in Fig. 2a the [1,2]- and [3,2]-migration transition states **6** and **7** are displaced along the homolytic coordinate axis relative to the ground states, at more or less constant acyloxy group charge.

Table 2 Activation barriers for 1,2-acyloxy migration in radicals **3a**–**e** through three membered ring transition state **6** and five membered ring transition state **7** (in kJ/mol), and cumulative charge and spin density values for the acyloxy groups as calculated at the UB3LYP/6-31G(d)// UB3LYP/6-31G(d) level of theory[14]

R_1	R_2	Level of theory	$\Delta E_0^{\ddagger}$ (6)	$\Delta E_0^{\ddagger}$ (7)	q/SD (3)	q/SD (6)	q/SD (7)	Reference
CH$_3$	H	UB3LYP/6-31G(d)	+77.0	+66.9	−0.34/0.07	−0.23/0.75	−0.30/0.56	17b
		QCISD/6-31G(d)	+104.2	+107.9				17b
CF$_3$	H	UB3LYP/6-31G(d)	+72.4	+58.6	−0.37/0.07	−0.32/0.65	−0.35/0.53	17b
		QCISD/6-31G(d)	+107.9	+98.3				17b
H	H	UMP2/6-31G(d)[a]		+97.1				4
		UB3LYP/6-31G(d)	+71.5	+57.3	−0.34/0.07	−0.25/0.73	−0.29/0.60	17b
		QCISD/6-31G(d)	+99.6	+98.7				17b
CH$_3$	CH$_3$	UB3LYP/6-31G(d)	+52.7	+47.3	−0.35/0.07	−0.35/0.58	−0.36/0.49	17c
		G3(MP2)B3	+72.0	+75.7				17c
CF$_3$	CH$_3$	UB3LYP/6-31G(d)	+42.7	+37.2	−0.39/0.07	−0.43/0.52	−0.43/0.44	17c
		G3(MP2)B3	+77.8	+65.7				17c

[a]Extrapolated value excluding differences in zero point vibrational energy.

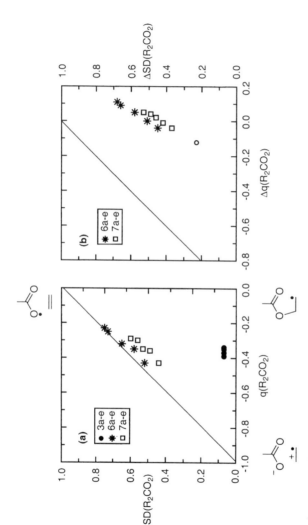

Fig. 2 (a) Graphical representation of the cumulative acyloxy group charge and spin densities for acyloxy radicals **3a–e** as calculated at the UB3LYP/6-31G(d) level of theory using the NPA scheme. (b) The same data as those in (a) represented as differences in charge and spin densities between ground states **3** and transition states **6** and **7**. The additional data point symbolized with an open circle corresponds to the difference between protonated systems **9** and **10**.[14]

Both reaction types should therefore be classified as mainly homolytic reaction types with little heterolytic character.[26] The dsdq-plot in Fig. 2b differs from the analogous presentation in Fig. 1b in that the horizontal Δq-axis runs from $+0.2$ to -0.8. The most interesting aspect of Fig. 2b is that all data points appear to line up along a line parallel to the diagonal, indicating an interrelation between Δq and ΔSD: the smaller the change in spin density located at the migrating group, the less positive/more negative is the migrating group and the lower is the activation barrier for the migration reaction.

With respect to the apparently dominant homolytic character as expressed in Fig. 2 it comes as a surprise that 1,2-acyloxy reactions appear to be acid catalyzed. A first indication for this possibility was obtained by comparing the reaction barriers calculated for the simple most model system **3c** ($R_1 = R_2 = H$) with that of its O-protonated form **9** (Scheme 5). The reaction barrier for the 1,2-acyloxy shift in **9** through three-membered ring transition state **10** has been calculated as $+15.1$ kJ/mol, which is 42.2 kJ/mol less than for the [3,2]-rearrangement through transition state **7c** and 56.4 kJ/mol less than for the [1,2]-rearrangement through transition state **6c** in the neutral parent system (B3LYP/6-31G(d) + ΔZPE results). The migrating acetate group carries a partial positive charge of $+0.47$ in ground state **9**, which is somewhat reduced to $+0.34$ in transition state **10**. The unpaired spin density located on the migrating group is at the same time increased from 0.11 in **9** to 0.33 in **10**.[14] Despite the fact that these charge and spin density data are rather different to those for the neutral uncatalyzed systems in absolute terms, the differences between ground state **9** and transition state **10** can be included in the dsdq-plot in Fig. 2b in order to attempt a comparison. It is interesting to see that the data point for the protonated system (open circle) falls onto the same correlation line observed for the neutral models. This implies that the mechanism respon- sible for synchronizing the changes in charge and spin densities in 1,2-acyloxy migration reactions is likely to be the same in the neutral as well as the cationic model systems.

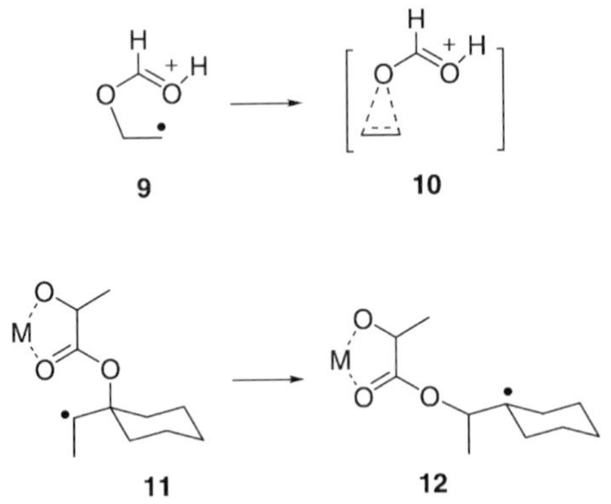

Scheme 5

Even though the magnitude of the proton-induced barrier reduction calculated here might not be attainable under solution phase conditions, this result still suggests that acid catalysis might play a significant role under some circumstances.

An experiment exploring the scope of Lewis acid catalysis in acyloxy rearrangement reactions has recently been published by Renauld et al. employing lactate ester radicals such as **11** with a variety of precoordinated Lewis acids M.[27] Rate accelerations of up to three orders of magnitude have been observed in these systems for the 1,2-acyloxy rearrangement with M = Sc(OTf)$_3$. The catalytic effects of (Lewis) acids present in the reaction mixture, whether added on purpose or present by accident, may also be the key to understanding some of the solvent effect data available for acyloxy rearrangement reactions. The rearrangement of **3d** has been found to be significantly accelerated by changing the solvent from tert-butyl benzene ($E_a = +75$ kJ/mol) to water ($E_a = +53$ kJ/mol).[28] However, systematically varying the solvent polarity between cyclohexane and methanol Beckwith et al. found only minor rate effects.[25] All efforts to reproduce the effects of aqueous solvation by theoretical modeling with either explicit or implicit solvation models[17b,c] have failed so far, indicating that either the solvation models are intrinsically incapable of estimating aqueous solvent effects for this reaction type or that the model systems studied lack some of the characteristics of the experimentally studied systems. One particular point of concern stems from the use of the TiCl$_3$/H$_2$O$_2$ couple in aqueous phase experiments to generate the radicals from closed shell precursors. This combination opens the possibility of either Lewis-acid catalysis through one of the titanium salts present in solution or even Brønstedt acid catalysis in a reaction mixture of uncontrolled acidity.

4   β-Phosphatoxyalkyl radicals

The chemistry of β-phosphatoxyalkyl radicals has been studied intensely in recent years due to the involvement of this structural motive in many biologically relevant organophosphate radicals.[3,29-33] In addition the premier leaving group abilities of the phosphate group has opened a non-oxidative route for the generation of alkene radical cations or CRIPs.[34] A first theoretical study on unimolecular reaction pathways in β-phosphatoxy radicals showed migration of the phosphatoxy group to be more facile than the acyloxy migration in an analogously substituted system, involving a larger preference for the [1,2]-migration pathway and a larger degree of charge separation.[35] The relative ease of 1,2-migration has also been demonstrated in experimental studies featuring both acyloxy and phosphatoxy groups in comparable positions.[36]

Theoretical studies have addressed the four reaction pathways shown in Scheme 6 including (i) the [1,2]-rearrangement reaction of reactant radical **13** through three-membered ring transition state **15** to yield product radical **14**, (ii) the analogous [3,2]-phosphatoxy rearrangement through five membered ring transition state **16**, (iii) heterolytic dissociation to yield the CRIP **17**, and (iv) the syn-1,3-elimination of

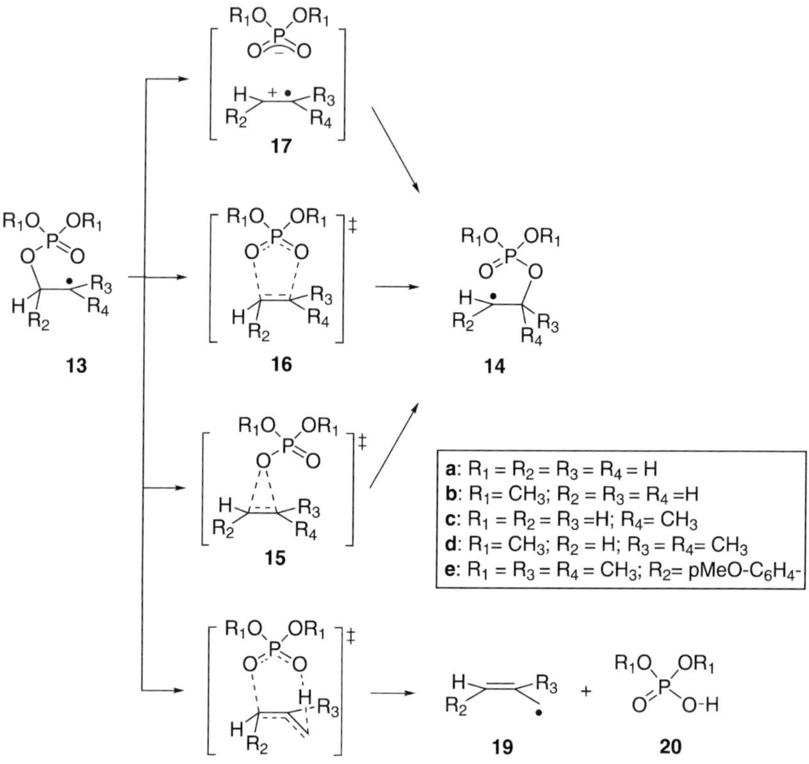

Scheme 6 structures:

Structure **17** (bracketed):
R_1O OR_1 / P / $O=\!=O$; $H{>}^{+}$ $\bullet{<}R_3$ / R_2 R_4

Structure **13**:
R_1O OR_1 / P / $O=$ O; $\bullet R_3$ / $H \diagup R_2$ R_4

Structure **16** (bracketed, ‡):
R_1O OR_1 / P / $O=$ O; R_3 R_4 / $H \diagup R_2$

Structure **14**:
R_1O OR_1 / P / $O=$ O; $H \cdot$ / R_2 R_3 R_4

Structure **15** (bracketed, ‡):
R_1O OR_1 / P / $O=$ O; H R_3 / R_2 R_4

Box:
a: $R_1 = R_2 = R_3 = R_4 = H$
b: $R_1 = CH_3$; $R_2 = R_3 = R_4 = H$
c: $R_1 = R_2 = R_3 = H$; $R_4 = CH_3$
d: $R_1 = CH_3$; $R_2 = H$; $R_3 = R_4 = CH_3$
e: $R_1 = R_3 = R_4 = CH_3$; $R_2 = pMeO\text{-}C_6H_4\text{-}$

Structure **18** (bracketed, ‡):
R_1O OR_1 / P / $O=$ O; H R_3 / $H \diagup R_2$

Structure **19**:
$H{>}{=}{<}R_3$ / R_2 $\bullet$

Structure **20**:
R_1O OR_1 / P / $O=$ $O\text{-}H$

Scheme 6

phosphoric acid **20** through transition state **18** yielding the allyl radical **19**. The last option is only valid for those systems in which either R_3 or R_4 are methyl groups.

　　The reaction barriers listed in Table 3 again provide some proof for the large dependence of the predicted reaction barrier on the level of theory. In particular the fraction of exact exchange contained in the hybrid functional appears to be critical, a larger fraction of exact exchange (as in the BHLYP functional) leading to higher barriers. Aside from these technical considerations the calculated barriers still allow some mechanistic conclusions to be drawn. The [1,2]-phosphatoxy rearrangement is clearly favored here over the competing [3,2]-alternative. The barrier for the *syn*-1,3-elimination pathway is predicted to be slightly higher in most systems than the most favorable rearrangement pathway. The *p*-methoxyphenyl substituted system **13e** is remarkable in several ways. First we note the very low barriers for practically all reaction pathways at B3LYP level. An experimental study of a closely related system ($R_1 = C_2H_5$) by Newcomb *et al.*[33e] sets the barrier for rearrangement in this system to $+40.6 \pm 5$ kJ/mol in THF. Additional consideration of solvent effects employing the PCM/UAHF solvent model[39] predicts solution phase barriers of

Table 3 Activation barriers for 1,2-phosphatoxy migration and *syn*-1,3-elimination in radicals **13a–e** (in kJ/mol)

System	Level of theory	$\Delta E_0^{\ddagger}$ (15)	$\Delta E_0^{\ddagger}$ (16)	$\Delta E_0^{\ddagger}$ (18)	Reference
13a	UB3LYP/6-31G(d)[a]	+79.9	+82.3		35
	UB3LYP/LB[b]	+71.5	+75.3		35
13b	UB3LYP/6-31G(d)[a]	+79.1	+84.5		35
	UB3LYP/LB[b]	+71.1	+79.5		35
13c	UB3LYP/6-31G(d)[a]	+71.1	+77.8	+85.8	35
	UB3LYP/LB[b]	+61.1	+68.6	+64.4	35
13d	UB3LYP/6-31G(d)[a]	+66.5	+77.4	+82.0	37
	UB3LYP/LB[b]	+56.5	+68.6	+61.5	37
13e	UB3LYP/6-31G(d)[a]	+33.0	+42.4	+38.3[c]	38
	UB3LYP/LB[b]	+23.9	+29.9	+26.0[c]	38
	UBHLYP/LB[d]	+52.4	+58.8	+61.0[c]	38

[a] ΔE_{tot} (UB3LYP/6-31G(d)) + ΔZPE (UB3LYP/6-31G(d)).
[b] ΔE_{tot} (UB3LYP/6-311 + G(d,p)) + ΔZPE(UB3LYP/6-31G(d)).
[c] Barriers for formation of CRIPs.
[d] ΔE_{tot} (UBHandHLYP/6-311 + G(d,p)) + ΔZPE(UB3LYP/6-31G(d)).

+17.6 kJ/mol for **15e** and +26.5 for **16e** at B3LYP/LB level and of +46.1 kJ/mol for **15e** and +55.4 for **16e** at BHLYP/LB level. Clearly the data predicted at this latter level are the only ones coming close to the experimentally measured ones. A second intriguing feature of system **13e** is that elimination of phosphoric acid now proceeds in a stepwise manner through initial formation of a CRIP structure. The barrier given in Table 3 for the elimination process is therefore identical to the barrier for formation of a CRIP intermediate. The BHLYP prediction for this process amounts to +61 kJ/mol in the gas phase and to +41.6 kJ/mol in THF solution. The energy of the actual CRIP **17e** is practically identical to the energy of transition state **18e**, indicating a minimal barrier for collapse of the CRIP towards reactant structure **13e** (and most likely also towards product structure **14e**). The actual lifetime of the CRIP intermediate **17e** might therefore be rather limited. The calculated activation barrier in THF as well as the minimal barrier for collapse are in direct support of the interpretation of the experimental results.

The charge and spin density distribution described in Table 4 for ground states **13** is hardly dependent on the substitution pattern and also rather similar to that in comparably substituted acyloxyalkyl radicals (Table 2). This implies that whatever appears as a substituent effect in the actual migration reactions cannot be a ground state effect. Solvent effects have only been explored for aryl substituted system **13e**. Compared to the substantial changes in reaction energetics the solvent induced changes in charge and spin density distribution are rather minor and lead to enhanced charge separation in the transition states. The most negatively charged phosphate groups in system **13e** occur along the [3,2]-migration pathway with gas

Table 4 Cumulative charges q and cumulative spin densities SD of the phosphatoxy groups in ground and transition states of phosphatoxy alkyl radicals **13a–13e** as calculated at the UB3LYP/6-31G(d)//UB3LYP/6-31G(d) level of theory[38,14]

System	13		15		16		18	
	q	SD	q	SD	q	SD	q	SD
13a	−0.35	0.03	−0.33	0.63	−0.42	0.37		
13b	−0.35	0.03	−0.32	0.65	−0.42	0.38		
13c	−0.37	0.07	−0.39	0.57	−0.47	0.34	−0.52	0.35
13d	−0.37	0.07	−0.50	0.32	−0.44	0.48	−0.54	0.31
13e	−0.37	0.06	−0.55	0.28	−0.70	0.20	−0.65	0.27
	−0.37[a]	0.06[a]	−0.58[a]	0.26[a]	−0.75[a]	0.16[a]	−0.72[a]	0.20[a]

[a]Charges in THF solution according to the PCM/UAF model.

and solution phase phosphate group charges of −0.70 and −0.75, respectively. This is rather close to what has been calculated for the CRIP structure **17e** that is accessible through transition state **18e** with phosphate group charges of −0.65 and −0.75 in the gas and THF solution phase, respectively. If the results obtained for **17e** are representative for CRIP complexes involving phosphate groups we may conclude that charge separation in CRIPs is not fully complete. This may, of course, severely affect their spectroscopic properties.

One interesting aspect of the sdq- and dsdq-plots in Fig. 3 describing phosphate group migration and elimination is the broad region covered by the systems described in Table 4. While the [1,2]-migration transition states **15a–c** appear to have mainly homolytic character, the transition states for [3,2]-migration **16e** and CRIP-formation **18e** have a strong heterolytic component. The high variability of the electronic characteristics of the various reaction pathways visible in Fig. 3 may be the root cause for the persistent debate over mechanistic details in the chemistry of β-phosphatoxy radicals.

5 β-Hydroxyalkyl radicals

The 1,2-migration of hydroxy groups in β-hydroxyalkyl radicals **21** has been studied repeatedly due to the involvement of these species in the enzyme-mediated dehydration reaction of 1,2-diols.[40–44] A detailed review of these results has recently been published by Radom *et al.*[45]

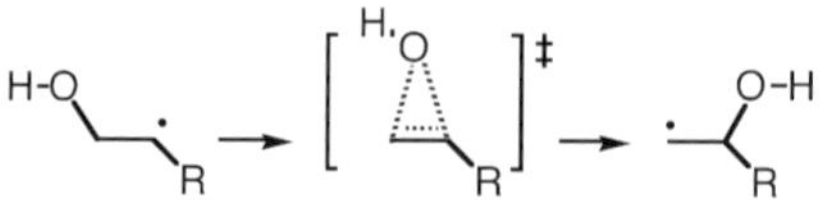

Scheme 7

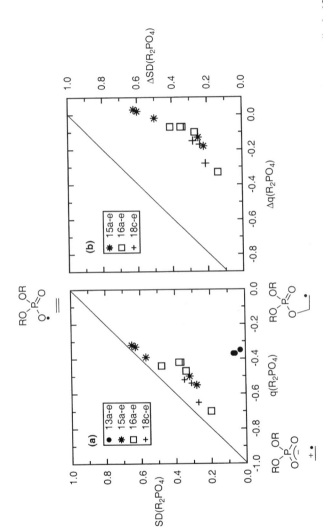

Fig. 3 (a) Graphical representation of the cumulative phosphatoxy group charge and spin densities for phosphatoxy radicals **13** and transition states **15**, **16** and **18** as calculated at the UB3LYP/6-31G(d) level of theory using the NPA scheme. (b) The same data as those in (a) represented as differences in charge and spin densities between ground states **13** and transition states **15**, **16** and **18**.[14]

While the 1,2-migration appears to be a high-barrier, stepwise process in the parent system (R = H), the reaction barrier is lowered on introduction of electron donating substituents R. For the ethylene glycol radical (R = OH) a concerted 1,2-migration pathway has been identified at a variety of theoretical levels and a barrier of +113 kJ/mol has been calculated at the G2(MP2,SVP)-RAD(p) level of theory. This is still only marginally less than the energy required for homolytic C–O bond dissociation to yield hydroxy radical OH· and vinyl alcohol and thus represents a rather unfavorable, high energy process. The consequences of acid catalysis have been studied extensively for this system and it has been found that even through partial protonation (i.e., complexation of the migrating hydroxy group to a good proton donor such as NH_4^+) the 1,2-migration barrier can be lowered to values that are in line with those estimated for the enzyme catalyzed process. This result is, in principle, analogous to the acid catalysis observed for the 1,2-migration process in β-acyloxyalkyl radicals (see Scheme 5). Coordination of the migrating hydroxy group to potassium cations has also been calculated to lower the 1,2-migration barrier.[46] One intriguing aspect of partial proton transfer catalysis with ammonium cations is the protonation state along the 1,2-migration pathway: while the proton resides on the ammonium catalyst in the reactant and product radicals 21 and 23, it has been transferred to the hydroxy group in the 1,2-migration transition state 22. This implies that transition state 22 is more basic than either reactant radical 21 or product radical 23 and may indeed hint at the intrinsically charge separating character of this rearrangement. Unfortunately, charge and spin density data appear not to be available for these systems.

6   β-Aminoalkyl radicals

The 1,2-migration of amino groups in β-aminoalkyl radicals such as 24 has also been of interest because of the involvement of these species in the enzyme-catalyzed elimination of ammonia from 1,2-amino alcohols.[47–50] These studies have also been reviewed recently by Radom et al.[45]

In close analogy to the 2-hydroxyethyl radical 21 (R = H) the 2-aminoethyl radical 24 faces a substantial barrier for the (most likely stepwise) 1,2-migration process. A transition state for the concerted migration pathway such as 25 could up to now not be located. What differentiates the amino 1,2-migration from the corresponding hydroxy group migration is that the former appears to be less affected

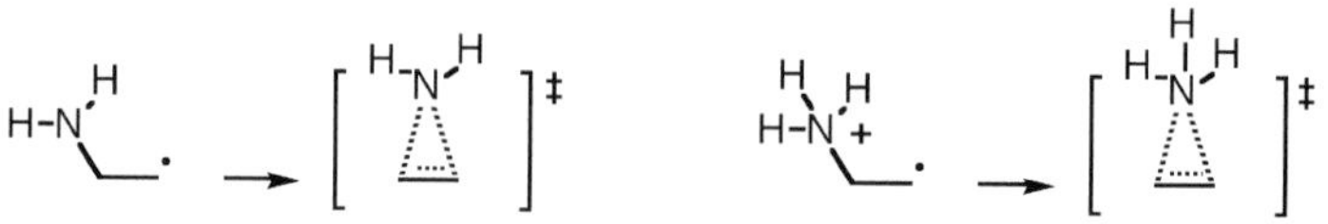

Scheme 8

by (partial) protonation of the migrating group. A barrier of $+104.8$ kJ/mol has been calculated for the 1,2-migration in **26** at the G2(MP2,SVP)-RAD(p) level of theory, the bridging structure **27** being a true transition state in this case.[49] Similar values have been obtained at other levels of theory.[48,50-52] These high barriers are somewhat in contrast to the frequent occurrence of the 1,2-amino group migrations in gas phase reactions of β-amino distonic radical cations such as **26**.[53,54] An analysis of the charge and spin density distribution has, unfortunately, not been performed in any of these reactions.

7 Conclusions

Concerted rearrangement or stepwise heterolytic dissociation/recombination? Considering the results collected in this account on a number of differently substituted systems the only acceptable answer to the key question posed at the beginning must be: it depends! It depends on the character of the migrating group, the substitution pattern, and the solvent polarity. It may also depend on the presence of a catalyst present in the reaction medium. That there is indeed not a single general mechanism for the 1,2-migration reaction in β-substituted alkyl radicals may most easily be illustrated with sdq- and dsdq-plots of four selected systems (Fig. 4).

This selection includes transition states from chlorine, acyloxy, and phosphatoxy migration reactions as well as three-and five-membered ring transition states in order to illustrate, how broadly the charge density/spin density space is covered in 1,2-migration reactions. That one and the same process can substantially change its characteristics as a function of solvent polarity and substitution pattern is, of course,

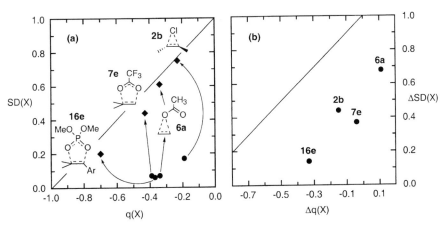

Fig. 4   (a) Graphical representation of migrating group (X) charge and spin densities for selected ground and transition states at the UB3LYP/6-31G(d) level of theory using the NPA scheme. (b) The same data as those in (a) represented as differences in charge and spin densities between ground and transition states.[14]

well known also from other reactions such as the large family of pericyclic reactions[1,55] or the S_N1/S_N2 mechanistic spectrum. While the term chameleonic may be used to describe this phenomenon in a compact fashion,[56] a conceptual basis can be found in the interplay of the dominant homolytic, heterolytic, and covalent VB configurations along the 1,2-migration pathway.[57]

Acknowledgements

This account is dedicated to Prof. K. N. Houk on occasion of his 60th birthday. He has been an inspiring teacher, a role model for computational chemists worldwide, and always a fun guy to have Sushi with.

References

1. Houk, K.N., Gonzalez, J. and Li, Y. (1995). *Acc. Chem. Res.* **28**, 81
2. Dewar, M.J.S. and Jie, C. (1992). *Acc. Chem. Res.* **25**, 537
3. Beckwith, A.L.J., Crich, D., Duggan, P.J. and Yao, Q. (1997). *Chem. Rev.* **97**, 3273
4. Saebo, S., Beckwith, A.L.J. and Radom, L. (1984). *J. Am. Chem. Soc.* **106**, 5119
5. Skell, P.S. and Traynham, J.G. (1984). *Acc. Chem. Res.* **17**, 160
6. (a) Hopkinson, A.C., Lieu, M.H. and Csizmadia, I.G. (1980). *Chem. Phys. Lett.* **71**, 557; (b) Hoz, T., Sprecher, M. and Basch, H. (1985). *J. Phys. Chem.* **89**, 1664; (c) Engels, B., Peyerimhoff, S.D. and Skell, P.S. (1990). *J. Phys. Chem.* **94**, 1267; (d) Guerra, M. (1992). *J. Am. Chem. Soc.* **114**, 2077; (e) Ihee, H., Zewail, A.H. and Goddard, W.A., III (1999). *J. Phys. Chem. A* **103**, 6638; (f) Zheng, X. and Phillips, D.L. (2000). *J. Phys. Chem. A* **104**, 1030; (g) Brana, P., Menendez, B., Fernandez, T. and Sordo, J.A. (2000). *J. Phys. Chem. A* **104**, 10842; (h) Li, Z.-H., Fan, K.-N. and Wong, M.W. (2001). *J. Phys. Chem. A* **105**, 10890; (i) Ihee, H., Kua, J., Goddard, W.A., III and Zewail, A.H. (2001). *J. Phys. Chem. A* **105**, 3623
7. Neumann, B. and Zipse, H. (2003). *Org. Biomol. Chem.* **1**, 168
8. Zipse, H., unpublished results
9. Reed, A.E., Curtiss, L.A. and Weinhold, F. (1988). *Chem. Rev.* **88**, 899
10. Engels, B. and Peyerimhoff, S.D. (1986). *J. Mol. Struct.* **138**, 59
11. Cao, J., Ihee, H. and Zewail, A.H. (1999). *Proc. Natl Acad. Sci. USA* **96**, 338
12. Ihee, H., Lobastov, V.A., Gomez, U., Goodson, B.M., Srinivasan, R., Ruan, C.-Y. and Zewail, A.H. (2001). *Science* **291**, 458
13. (a) More O'Ferrall, R.A. (1970). *J. Chem. Soc.* 274; (b) Jencks, W.P. (1972). *Chem. Rev.* **72**, 706
14. All charge and spin density values mentioned in this account have been calculated using the Natural Population Analysis scheme and the Becke3LYP hybrid functional as implementation in GAUSSIAN 98, Rev. A.11, together with the pruned ultrafine grid of 99 radial shells and 590 angular points per atom as well as a tight conversion criterion. All values are given in atomic units
15. (a) Tan, E.W. and Shaw, J.P. (1996). *J. Org. Chem.* **61**, 5635; (b) Tan, E.W., Chan, B. and Blackman, A.G. (2002). *J. Am. Chem. Soc.* **124**, 2078
16. Cozens, F.L., O'Neill, M., Bogdanova, R. and Schepp, N. (1997). *J. Am. Chem. Soc.* **116**, 10652

17. (a) Zipse, H. (1996). *J. Chem. Soc. Perkin Trans.* **2**, 1797; (b) Zipse, H. (1997). *J. Am. Chem. Soc.* **119**, 1087; (c) Zipse, H. and Bootz, M. (2001). *J. Chem. Soc. Perkin Trans.* **2**, 1566

18. (a) Beckwith, A.L.J. and Tindal, P.L. (1971). *Aust. J. Chem.* **24**, 2099; (b) Perkins, M.J. and Roberts, B.P.J. (1975). *J. Chem. Soc. Perkin Trans.* **2**, 77

19. Barclay, L.R.C., Griller, D. and Ingold, K.U. (1982). *J. Am. Chem. Soc.* **104**, 4399

20. Kocovsky, P., Stary, I. and Turecek, F. (1986). *Tetrahedron Lett.* **27**, 1513

21. Korth, H.-G., Sustmann, R., Gröninger, K.S., Leising, M. and Giese, B. (1988). *J. Org. Chem.* **53**, 4364

22. Beckwith, A.L.J. and Duggan, P.J. (1992). *J. Chem. Soc. Perkin Trans.* **2**, 1777

23. Beckwith, A.L.J. and Duggan, P.J. (1993). *J. Chem. Soc. Perkin Trans.* **2**, 1673

24. Crich, D. and Filzen, G.F. (1995). *J. Org. Chem.* **60**, 4834

25. Beckwith, A.L.J. and Duggan, P.J. (1996). *J. Am. Chem. Soc.* **118**, 12838

26. Zipse, H. (1999). *Acc. Chem. Res.* **32**, 571

27. Lancote, E. and Renaud, P. (1998). *Angew. Chem.* **110**, 2369 *Angew. Chem. Int. Ed.* 1998, **37**, 2259

28. Barclay, L.R.C., Lusztyk, J. and Ingold, K.U. (1984). *J. Am. Chem. Soc.* **106**, 1793

29. Knapp-Pogozelski, W. and Tullius, T.D. (1998). *Chem. Rev.* **98**, 1089

30. Breen, A.P. and Murphy, J.A. (1995). *Free Rad. Biol. Med.* **18**, 1032

31. (a) Dizdaroglu, M., von Sonntag, C. and Schulte-Frohlinde, D. (1975). *J. Am. Chem. Soc.* **97**, 2277; (b) Behrens, G., Koltzenburg, G., Ritter, A. and Schulte-Frohlinde, D. (1978). *Int. J. Rad. Biol.* **33**, 163; (c) Koltzenburg, G., Behrens, G. and Schulte-Frohlinde, D. (1982). *J. Am. Chem. Soc.* **104**, 7311

32. (a) Glatthaar, R., Spichty, M., Gugger, A., Batra, R., Damm, W., Mohr, M., Zipse, H. and Giese, B. (2000). *Tetrahedron* **56**, 4117; (b) Meggers, E., Dussy, A., Schäfer, T. and Giese, B. (2000). *Chem. Eur. J.* **6**, 485; (c) Gugger, A., Batra, R., Rzadek, P., Rist, G. and Giese, B. (1997). *J. Am. Chem. Soc.* **119**, 8740; (d) Giese, B., Beyrich-Graf, X., Erdmann, P., Petretta, M. and Schwitter, U. (1995). *Chem. Biol.* **2**, 367; (e) Giese, B., Beyrich-Graf, X., Burger, J., Kesselheim, C., Senn, M. and Schäfer, T. (1993). *Angew. Chem. Int. Ed. Engl.* **32**, 1742

33. (a) Newcomb, M., Miranda, N., Sannigrahi, M., Huang, X. and Crich, D. (2001). *J. Am. Chem. Soc.* **123**, 6445; (b) Horner, J.H. and Newcomb, M. (2001). *J. Am. Chem. Soc.* **123**, 4364; (c) Bales, B.C., Horner, J.H., Huang, X., Newcomb, M., Crich, D. and Greenberg, M.M. (2001). *J. Am. Chem. Soc.* **123**, 3623; (d) Newcomb, M., Miranda, N., Huang, X. and Crich, D. (2000). *J. Am. Chem. Soc.* **122**, 6128; (e) Whitted, P.O., Horner, J.A., Newcomb, M., Huang, X. and Crich, D. (1999). *Org. Lett.* **1**, 153; (f) Newcomb, M., Horner, J.H., Whitted, P.O., Crich, D., Huang, X., Yao, Q. and Zipse, H. (1999). *J. Am. Chem. Soc.* **121**, 10685; (g) Choi, S.-Y., Crich, D., Horner, J.H., Huang, X., Martinez, F.N., Newcomb, M., Wink, D.J. and Yao, Q. (1998). *J. Am. Chem. Soc.* **120**, 211

34. (a) Crich, D. and Ranganathan, K. (2002). *J. Am. Chem. Soc.* **124**, 12422; (b) Crich, D. and Neelamkavil, S. (2002). *Org. Lett.* **4**, 2573; (c) Crich, D., Ranganathan, K. and Huang, X. (2001). *Org. Lett.* **3**, 1917; (d) Crich, D., Huang, X. and Newcomb, M. (2000). *J. Org. Chem.* **65**, 523; (e) Crich, D., Huang, X. and Newcomb, M. (1999). *Org. Lett.* **1**, 225

35. Zipse, H. (1997). *J. Am. Chem. Soc.* **119**, 2889

36. Müller, S.N., Batra, R., Senn, M., Giese, B., Kisel, M. and Shadyro, O. (1997). *J. Am. Chem. Soc.* **119**, 2796

37. Newcomb, M., Horner, J.H., Whitted, P.O., Crich, D., Huang, X., Yao, Q. and Zipse, H. (1999). *J. Am. Chem. Soc.* **121**, 10685

38. Wang, Y. and Zipse, H., Submitted for publication

39. (a) Barone, V., Cossi, M. and Tomasi, J. (1997). *J. Chem. Phys.* **107**, 3210; (b) Barone, V. and Cossi, M. (1998). *J. Phys. Chem. A* **102**, 1995; (c) Amovilli, C., Barone, V., Cammi,

R., Cances, E., Cossi, M., Mennucci, B., Pomelli, C.S. and Tomasi, J. (1998). *Adv. Quant. Chem.* **32**, 227

40. Smith, D.M., Golding, B.T. and Radom, L. (2001). *J. Am. Chem. Soc.* **123**, 1664
41. Smith, D.M., Golding, B.T. and Radom, L. (1999). *J. Am. Chem. Soc.* **121**, 5700
42. George, P., Glusker, J.P. and Bock, Ch.W. (1997). *J. Am. Chem. Soc.* **119**, 7065
43. George, P., Glusker, J.P. and Bock, Ch.W. (1995). *J. Am. Chem. Soc.* **117**, 10131
44. Golding, B.T. and Radom, L. (1976). *J. Am. Chem. Soc.* **98**, 6331
45. Smith, D.M., Wetmore, S.D. and Radom, L. (2001). In *Theoretical Biochemistry – Processes and Properties of Biological Systems*, Ericksson, L.A. (ed.), pp. 183–214. Elsevier, Amsterdam
46. (a) Toraya, T., Yoshizawa, K., Eda, M. and Yamabe, T. (1999). *J. Biochem.* **126**, 650; (b) Eda, M., Kamachi, T., Yoshizawa, K. and Toraya, T. (2002). *Bull. Chem. Soc. Jpn* **75**, 1469
47. Wetmore, S.D., Smith, D.M., Bennett, J.T. and Radom, L. (2002). *J. Am. Chem. Soc.* **124**, 14054
48. Semialjac, M. and Schwarz, H. (2002). *J. Am. Chem. Soc.* **124**, 8974
49. Wetmore, S.D., Smith, D.M. and Radom, L. (2001). *J. Am. Chem. Soc.* **123**, 8678
50. Wetmore, S.D., Smith, D.M. and Radom, L. (2000). *J. Am. Chem. Soc.* **122**, 10208
51. Hammerum, S. (2000). *Int. J. Mass Spectrom.* **199**, 71
52. Yates, B.F. and Radom, L. (1987). *Org. Mass Spectrom.* **22**, 430
53. Hammerum, S., Petersen, A.C., Sølling, T.I., Vulpius, T. and Zappey, H. (1997). *J. Chem. Soc. Perkin Trans.* **2**, 391
54. Bjornholm, T., Hammerum, S. and Kuck, D. (1988). *J. Am. Chem. Soc.* **110**, 3862
55. Houk, K.N., Li, Y. and Evanseck, J.D. (1992). *Angew. Chem. Int. Ed. Engl.* **31**, 682
56. Doering, W.v.E. and Wang, Y. (1999). *J. Am. Chem. Soc.* **121**, 10112
57. (a) Shaik, S. and Shurki, A. (1999). *Angew. Chem. Int. Ed. Engl.* **38**, 586; (b) Shaik, S.S., Schlegel, H.B. and Wolfe, S. (1992). *Theoretical Aspects of Physical Organic Chemistry. Application to the S_N2 Transition State*. Wiley Interscience, New York

Computational studies of alkene oxidation reactions by metal-oxo compounds[☆]

THOMAS STRASSNER

Technische Universität München, Anorganisch-chemisches Institut, Lichtenbergstraße 4, D-85747 Garching bei München, Germany

Dedicated to Ken Houk

1 Introduction

Stoichiometric and catalytic transition-metal oxidation reactions are of great interest, because of their important role in industrial and synthetic processes. The oxidation of alkenes is one of the fundamental reactions in chemistry.[1] Most bulk organic products contain functional groups, which are produced in the chemical industry by direct oxidation of the hydrocarbon feedstock. Usually these reactions employ catalysts to improve the yields, to reduce the necessary activation energy and render the reaction more economic. The synthesis of almost every product in chemical industry nowadays employs at least one catalytic step. The oxidation products of alkenes, epoxides and glycols, may be transformed into a variety of functional groups and therefore the selective and catalytic oxidation of alkenes is an industrially important process.

Several oxidants (Fig. 1) are used as the oxygen source. Examples are bleach (NaOCl), hydrogen peroxide (H$_2$O$_2$), organic peroxides like dimethyldioxyrane (DMD) or *tert*-butyl hydroperoxide (TBHP), peracids like *m*-chloroperbenzoic acid (*m*CPBA) or potassium monoperoxysulfate (KHSO$_5$).

[☆] Supporting information for this article is available from the author.

ADVANCES IN PHYSICAL ORGANIC CHEMISTRY
VOLUME 38 ISSN 0065-3160   DOI 10.1016/S0065-3160(03)38004-9

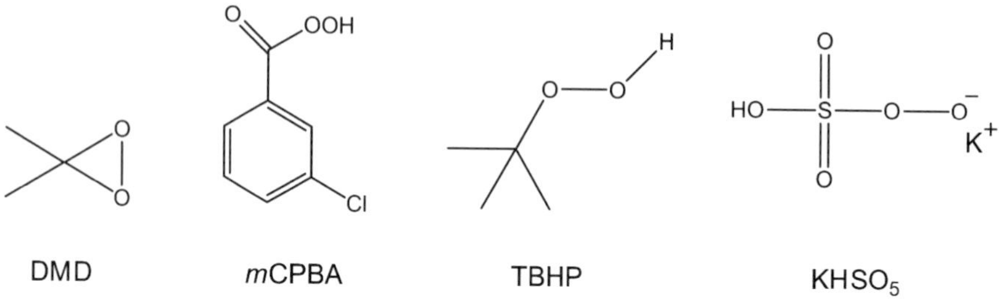

Fig. 1 Examples of oxygen sources.

But for economic reasons oxygen or even air is the most attractive oxidant for the chemical industry.

The activation of oxygen in oxygen transfer reactions is usually mediated by a suitable transition metal catalyst which has to be sufficiently stable under the reaction conditions needed. But also non-metal catalysts for homogeneous oxidations have recently been of broad interest and several of them have been compiled in a recent review.[2] Other examples for well known alkene oxidation reactions are the ozonolysis, hydroboration reactions or all biological processes, where oxygen is activated and transferred to the substrate. Examples for these reactions might be cytochrome P_{450} or other oxotransferases. Of these reactions, this contribution will focus on transition-metal mediated epoxidation and dihydroxylation.

Scheme 1 shows the three pathways which have been found in the cases described below. The epoxidation pathway proceeds either by formation of a metal-peroxo species or direct transfer of the oxygen, while in the case of the dihydroxylation the transfer of the oxygen proceeds via a concerted process.

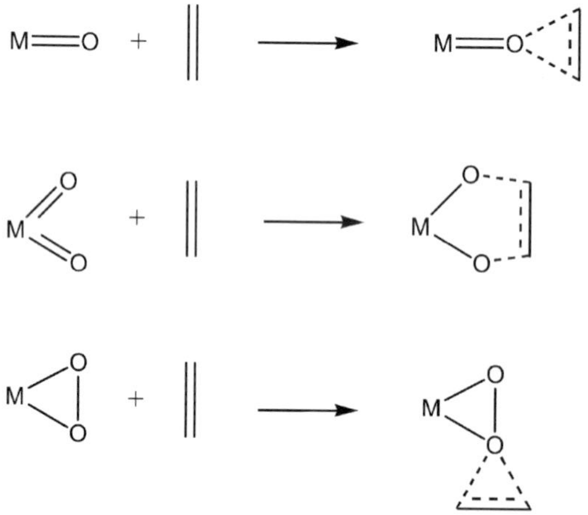

Scheme 1 Proposed transition states for the interaction of metal-oxo compounds with alkenes.

Theoretical investigations have become more and more important for the development of new catalysts. The interaction of experimental and quantum chemists is fruitful because of the better accuracy and the possibility to calculate the molecules instead of "model systems." Quantum-chemical calculations now allow for the determination of transition state structures and an analysis of the factors which have an impact on the reaction.

The enormous progress in computational techniques during the last years is reflected by the higher level applications of *ab initio* Hartree–Fock (HF), post-HF and density functional theory calculations (DFT). DFT calculations have been shown to be superior to HF or post-HF methods for the treatment of transition metals and are generally accepted as the best method for the calculation of catalyst systems containing transition metals.[3-6] They allow the prediction of important chemical and physical properties of the metal complexes involved in these reactions.[7] In concert with the increasing computational power of modern computers and improvements of the quantum chemistry codes and algorithms, it is now possible to examine "real" problems that could not be tackled earlier.[8] The accuracy of the results for transition metal compounds is nowadays as good as what has been achieved earlier only for small organic molecules.[9]

The resolution of the hot debate on the mechanism of metal-oxo mediated oxidations is one of the success stories of DFT calculations. An early publication by Sharpless on chromylchloride oxidations of alkenes[10] started a long ongoing discussion[11-25] on the mechanism of metal-oxo mediated oxidations. Sharpless proposed an interaction between the chromium metal and the alkene and generalized his proposal to include all metal-oxo compounds, especially osmium tetroxide and permanganate. Especially the mechanism of the reaction of osmium tetroxide with alkenes was the subject of an intense debate within the community of experimental organic chemists (Scheme 2).

It was generally accepted that the reaction proceeds via a concerted mechanism with a cyclic ester intermediate (Scheme 2, [3 + 2]), until Sharpless suggested the stepwise mechanism[10] via a metallaoxetane intermediate (Scheme 2, [2 + 2]), which is supposed to rearrange (Scheme 2, RA) to a cyclic ester before the hydrolysis takes place.

The quotes given below illustrate the nature of the disagreement. It was not possible to distinguish between the two main proposals until this controversy was addressed by several high-level theoretical studies[26-29]:

"...a concerted mechanism between OsO_4 and olefins via a [3 + 2] pathway is proposed", R. Criegee, *Justus Liebigs Ann. Chem.* **1936**, *522*, 75.

"...we propose an alternative mechanism which involves a four-membered organoosmium intermediate...", K.B. Sharpless *et al.*, *J. Am. Chem. Soc.* **1977**, *99*, 3120.

"...the frontier orbitals in osmium tetroxide are set up for a [3 + 2] reaction, whereas a geometric distortion of osmium tetroxide would have to take place if

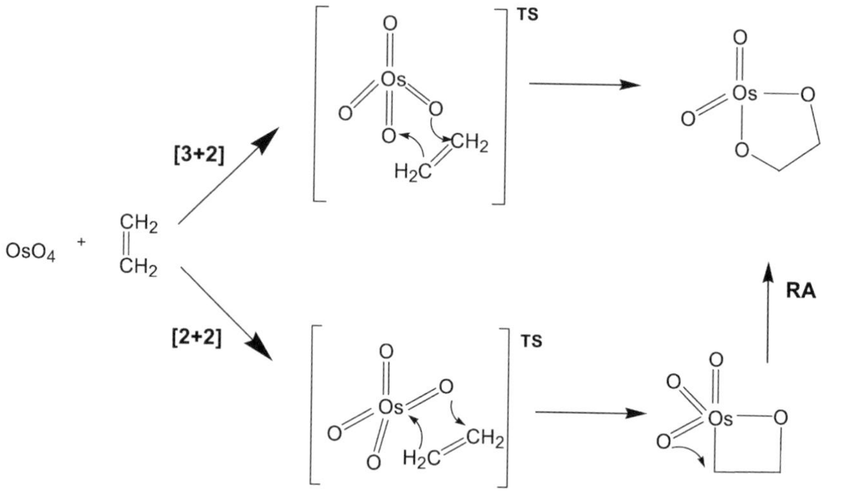

Scheme 2 Mechanistic proposals for the osmium tetraoxide oxidation of alkenes (RA = rearrangement): [3 + 2]- vs. stepwise [2 + 2]-reaction.

the reaction were a [2 + 2] followed by a second deformation…attention will be focused mainly on a concerted mechanism, although one cannot rule out on theoretical grounds the mechanism proposed by Sharpless" K.A. Jørgensen, R. Hoffmann, *J. Am. Chem. Soc.* **1986**, *108*, 1867.

"… we report new data which allow the [2 + 2] pathway to be excluded from consideration…", E.J. Corey *et al.*, *J. Am. Chem. Soc.* **1993**, *115*, 12579.

"…our model provides reasonable structures for both [3 + 2] and [2 + 2]…, so contrary to their claim, the data presented does not exclude a [2 + 2] mechanism involving an osmaoxetane…", K.B. Sharpless, *Tetrahedron Lett.* **1994**, *35*, 7315.

"Temperature effects in assymetric dihydroxylation – evidence for a stepwise [2 + 2] mechanism", Sharpless *et al.*, *Angew. Chem. Int. Ed. Engl.* **1993**, *32*, 1329.

"…it was not possible to reconcile much of the experimental evidence with a metallaoxetane-like transition state..", E.J. Corey *et al.*, *Tetrahedron Lett.* **1996**, *37*, 4899.

"…the results to date indicate that the AD proceeds…by a pathway which is most consistent with the ligated osmaoxetane intermediate previously proposed", Sharpless *et al.*, *J. Am. Chem. Soc.* **1997**, *119*, 1840.

"…it should be noted that the [2 + 2] osmaoxetane pathway is inconsistent with the observed absolute stereocourse of the reaction…", E.J. Corey *et al.*, *J. Am. Chem. Soc.* **1996**, *118*, 7851.

"Though the mechanism of even the "simple" reaction of osmium tetroxide with an olefin remains uncertain, there has been considerable controversy over the mechanism of the process when a chiral cinchona alkaloid ligand is also involved". K.B. Sharpless *et al.*, *J. Org. Chem.* **1996**, *61*, 7978.

The results of the DFT-calculations on the osmium tetroxide oxidation of alkenes are described in detail in Section 2.1.

Many groups have started to undertake investigations to draw distinctions between reaction mechanisms which most of the time cannot be distinguished by experimental studies. The description of such investigations in this chapter cannot be comprehensive, but is restricted to several examples. I apologize to the authors of other important papers in this field which are neither mentioned nor at least cited in this review.

2 Dihydroxylation

The introduction of oxygen atoms into unsaturated organic molecules via dihydroxylation reactions leading to 1,2-diols is an important reaction. 1,2-Diols can be synthesized by the reaction of alkenes with organic peracids via the corresponding epoxides and subsequent hydrolysis or metal-catalyzed oxidation by strong oxidants such as osmium tetroxide (OsO_4), ruthenium tetroxide (RuO_4) or permanganate (MnO_4^-) to name only a few. The dihydroxylation reaction proceeds in the first reaction step via the [3 + 2] pathway forming a dioxylate and was investigated in detail for MO_3^q and LMO_3^q ($q = 1, 0, -1$) systems for different ligands L (cp, Cl^-, CH_3, O) by Rösch.[30] They reported reaction energies, activation barriers and transition states, which were correlated using Marcus theory as well as with the M–O bond dissociation energies (BDE) of the reactants. The observed correlation can be used not only to predict the reaction energy from the BDE of similar complexes, but also to estimate the activation barriers via the Marcus equation.[30] Generally the reactions of complexes LMO_3^q ($q = 1, 0, -1$) show a lower BDE than those of the corresponding MO_3^q ($q = 1, 0, -1$) systems, a greater reaction exothermicity and lower activation barriers.

Osmium tetroxide and permanganate are the textbook examples for the direct addition of the hydroxyl function to double bonds as shown in Scheme 3. They have been rationalized to be feasible because of their large thermodynamic exothermicities,[30] and the existence of a low-energy pathway discussed in Section 2.1 for the transfer of two oxygen atoms from the metal to the adjacent alkene carbons.

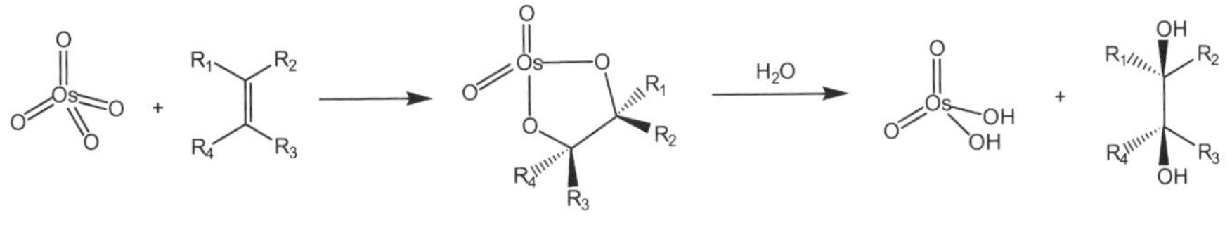

Scheme 3 *Cis*-dihydroxylation reaction of alkenes by osmiumtetraoxide.

OSMIUM TETROXIDE OSO$_4$

Quite recently it was reported that in addition to hydrogen peroxide, periodate or hexacyanoferrat(III), molecular oxygen[21,31–34] can be used to reoxidize these metal-oxo compounds. New chiral centers in the products can be created with high enantioselectivity in the dihydroxylation reactions of prochiral alkenes. The development of the catalytic asymmetric version of the alkene dihydroxylation was recognized by Sharpless' receipt of the 2001 Nobel prize in Chemistry.

The reaction mechanisms shown in Scheme 2 have been the subject of several computational studies. Of particular interest has been the dihydroxylation by osmium tetroxide,[26–29] where the above mentioned controversy about the mechanism of the oxidation reaction with olefins could not be solved experimentally.[10,12,13,16,19,22,24,25,35,36]

Sharpless[10] proposed a stepwise [2 + 2] mechanism based on the partial charges of metal and oxygen atoms and concluded, that metallaoxetanes should be involved in alkene oxidation reactions of metal-oxo compounds like CrO_2Cl_2, OsO_4 and MnO_4^-. The question arose whether the reaction proceeds via a concerted [3 + 2] route as originally proposed by Criegee[35,37] or via a stepwise [2 + 2] process with a metallaoxetane intermediate[10] (Scheme 2).

As early as 1936 Criegee observed that the rate of this reaction increases when bases such as pyridine are added. Kinetic data on the influence of the reaction temperature on the enantioselectivity of the dihydroxylation of prochiral alkenes in the presence of chiral amines revealed a non-linearity of the modified Eyring plot.[25] The deviation from linearity and the existence of an inversion point in this plot indicated that two different transition states are involved, inconsistent with a concerted [3 + 2] mechanism.

Sharpless also found that chirality can be transferred to the substrates by chinchona amines, which led to the development of the asymmetric version of the reaction.

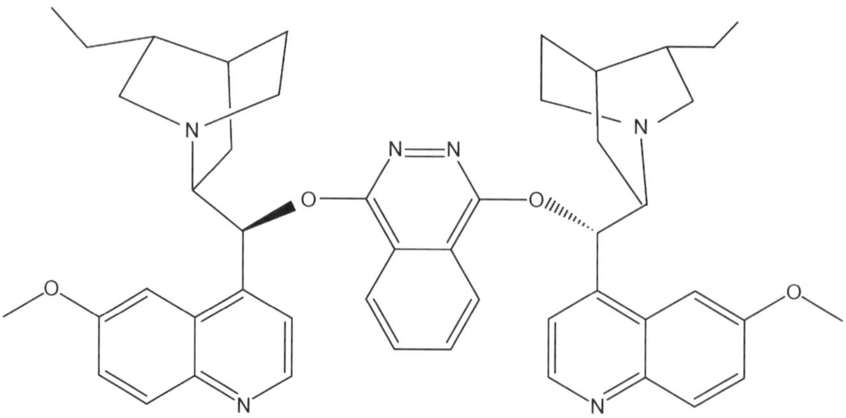

Fig. 2   Chinchona-Base (DHQ)$_2$PHAL.

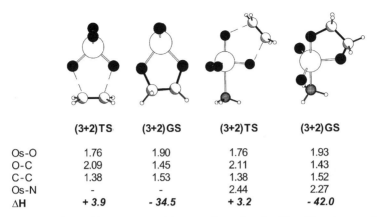

	(3+2)TS	(3+2)GS	(3+2)TS	(3+2)GS
Os-O	1.76	1.90	1.76	1.93
O-C	2.09	1.45	2.11	1.43
C-C	1.38	1.53	1.38	1.52
Os-N	-	-	2.44	2.27
ΔH	+ 3.9	- 34.5	+ 3.2	- 42.0

Fig. 3 Calculated transition states and intermediates for the [3 + 2]-pathway with and without base[29] (bond length in Å, enthalpy in kcal/mol).

The chinchona bases lead to high enantiomeric excesses and are part of the commercially available AD-mix (0.4% $K_2OsO_2(OH)_4$, 1.0% $(DHQ)_2PHAL$, 300% $[K_3Fe(CN)_6]$, 300% K_2CO_3).

The computational power has increased significantly during the last years, but it is still not feasible to evaluate a potential energy surface and to optimize complexes with bases as large as the chinchona bases. In the DFT studies[26–29] described here the large bases have been modeled by NH_3.

A comparison of the computed transition state structures for reactions in the presence and absence of basic ligand shows no large structural differences (Figs. 3 and 4). Table 1 shows that the differences in the enthalpy of activation for these reactions are less than 1 kcal/mol and that the [3 + 2] pathway is significantly lower in energy compared to the [2 + 2] pathway.

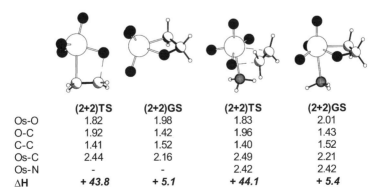

	(2+2)TS	(2+2)GS	(2+2)TS	(2+2)GS
Os-O	1.82	1.98	1.83	2.01
O-C	1.92	1.42	1.96	1.43
C-C	1.41	1.52	1.40	1.52
Os-C	2.44	2.16	2.49	2.21
Os-N	-	-	2.42	2.42
ΔH	+ 43.8	+ 5.1	+ 44.1	+ 5.4

Fig. 4 Calculated transition states and intermediates for the [2 + 2]-pathway with and without base[29] (bond length in Å, enthalpy in kcal/mol).

Table 1 Base-catalyzed dihydroxylation reaction

	(3 + 2)TS	(3 + 2)GS	(2 + 2)TS	(2 + 2)GS
Houk[29]	+3.2	−42.0	+44.1	+5.4
Frenking[26]	+4.4	−39.8	+44.3	+5.1
Morokuma[27]	+1.4	−23.5	+50.4	+13.1
Ziegler[28]	+0.8	−28.4	+39.1	+3.6

In just a short period of time four different groups published the results of DFT studies on this reaction using different quantum chemical packages and levels of theory. It was concluded in each study that the barrier for the [2 + 2] addition of OsO_4 to ethylene and the ring expansion are significantly higher (~35 kcal/mol) than the activation enthalpy needed for the [3 + 2]) pathway.

A combined experimental and theoretical study using kinetic isotope effects (KIEs) to compare experiment and theory provided additional evidence that the reaction proceeds via a [3 + 2] pathway. The KIEs were measured by a new NMR technique[38] and were compared to values, which can be obtained from the calculated transition state structures. Two sets of data were measured for the experimentally used alkene $(H_3C)_3C–CH=CH_2$ (Scheme 4), and the structures of the corresponding transition states for substituted alkenes were also calculated.

Propene was chosen as the model system for these calculations. The number of possible transition states rose significantly when all possible orientations were calculated. Figure 5 shows six different transition states for the [2 + 2] pathway with propene.

Similarly, several [3 + 2] transition states were identified together with transition states for the rearrangement reaction. The calculation of KIEs was undertaken for the low energy transition states. As an example, the two [3 + 2] transition states with the lowest activation barriers are shown in Fig. 6, together with a comparison between the calculated KIE for the given transition state structures and the experimentally determined KIE.

The experimental and theoretical values for the KIEs were found to match only in the case of the [3 + 2] pathway (Fig. 6) and it was concluded that, indeed, only the [3 + 2] pathway is feasible.[29]

The disadvantage of using NH_3 as a model for the chinchona bases is the failure to account for the steric effects of the bulky amine base. Therefore QM/MM-calculations[39,40] were carried out. These combine the advantages of high level QM calculations for a small fragment, with the treatment of a much larger number of

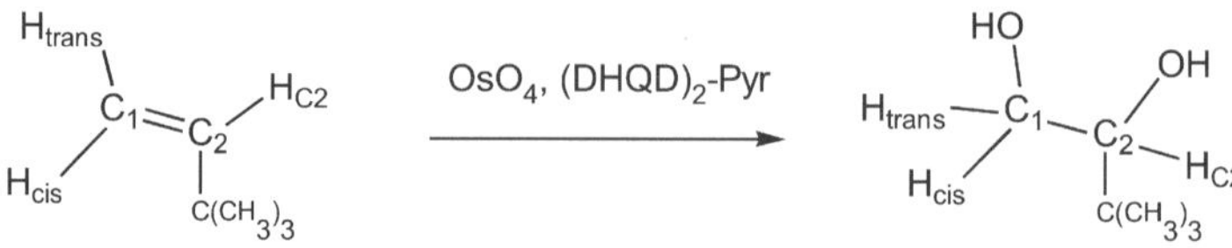

Scheme 4 Kinetic isotope effect experiment.[29]

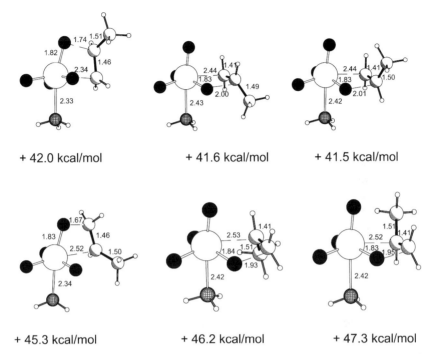

Fig. 5 Optimized transition states for the reaction of osmium tetroxide with propene via a [2 + 2]-pathway.

atoms by molecular mechanics.[41] The origin of the enantioselectivity observed in the dihydroxylation of styrene was investigated and it was found that this is the result of π-interactions between the aromatic rings of the reactant and catalyst. Norrby[40] has parametrized a force field for this reaction and used it to reproduce the experimentally observed enantioselectivities. Several examples are shown in Table 2.

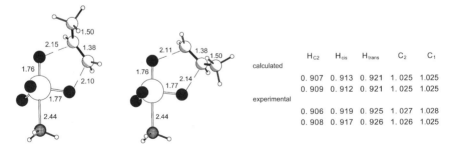

Fig. 6 Calculated [3 + 2]-transition states and a comparison of their KIEs to experimentally measures values (Scheme 4).

Table 2 Calculated and experimental enantioselectivities in the AD (adapted from Ref. 40)

Alkene	DHQD ligand	ee_{calc} (%)	ee_{exp} (%)	Reference
1-Phenyl-cyclohexene	CLB	91	91	42
Styrene	CLB	70	74	42
β, β-Dimethyl styrene	CLB	72	74	42
β-Vinyl naphthalene	CLB	94	88	42
trans-Stilbene	CLB	98	99	42
t-Butyl ethene	CLB	70	44	40
α-Methyl styrene	CLB	65	62	40
Styrene	MEQ	94	87	42
Styrene	PHN	98	78	42
t-Butyl ethene	PHN	89	79	42
β-Vinyl naphthalene	PHAL	100	98	20
Styrene	PHAL	97	97	43
α-Methyl styrene	PHAL	99	94	43
trans-Stilbene	PHAL	100	100	43

PERMANGANATE (MNO_4^-)

The oxidation of alkenes by permanganate is one of the frequently used examples in freshman chemistry. It is also well known as the Baeyer test for unsaturation. There are many reagents that add two hydroxyl groups to a double bond,[44] but osmium tetroxide and permanganate are the most prominent ones. The mechanism of the permanganate oxidation is believed to be similar to the oxidation of alkenes by OsO_4.[45,46]

It was generally accepted that the reaction proceeds via a concerted mechanism with a cyclic ester intermediate, until the suggestion was made that this reaction also might proceed by a stepwise mechanism through a metallaoxetane intermediate,[10] which then rearranges to a cyclic ester intermediate that undergoes hydrolysis to form the final diol. Just as for osmium tetroxide (Scheme 2), the proposed [2 + 2]- and [3 + 2]-pathways could not be distinguished on the basis of experimental data.

Until now several groups have failed to identify the elusive metallaoxetane and the extensive set of available kinetic data provide no indication for the existence of the species. But the possibility that it might be a non-rate-determining intermediate could not be excluded experimentally. Different mechanisms were proposed to explain the variety of experimental results available, but the mechanistic issues remain unresolved.

DFT-calculations show great similarities between the alkene dihydroxylation by permanganate and osmium tetroxide. The activation energy for the [3 + 2]-pathway is a little higher in energy (+9.2 kcal/mol) compared to osmium tetroxide, while the barrier for the [2 + 2]-pathway is more than 40 kcal/mol higher in energy (+50.5 kcal/mol).[47]

The calculated structures of the transition states for these two pathways for the oxidation of ethylene are shown in Fig. 7.

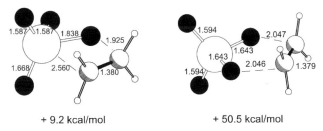

+ 9.2 kcal/mol   + 50.5 kcal/mol

Fig. 7 [2 + 2]- and [3 + 2]-transition state for the permanganate oxidation of alkenes.[47]

Secondary kinetic isotope effects of $k_H/k_D = 0.77$ (α-H) and $k_H/k_D = 0.75$ (β-H) were determined by experiment for the oxidative cleavage of cinnamic acid by acidic permanganate.[48] A later paper from the same group on the same reaction using quaternary ammonium permanganates[49] reported very different isotope effects of $k_H/k_D = 1.0$ (α-H) and $k_H/k_D = 0.91-0.94$ (β-H) depending on the counter-ion. B3LYP/6-311+G** calculations reproduced very well the reported activation energy of 4.2 ± 0.5 kcal/mol[50] with a predicted activation enthalpy of 5.1 kcal/mol for the transition state shown in Fig. 8. It is obvious that the expected secondary isotope effects at the two alkene carbon atoms must be very different, because of the large difference in the two C–O bond lengths at the transition state (1.94 and 2.20 Å, respectively). Additionally two sets of ^{13}C KIEs were determined experimentally by the same method that had been used for the investigation of the osmium tetroxide reaction. Comparing these to theoretically predicted values they agree very well within the experimental uncertainty.[51]

These results indicate that also in the case of permanganate the dihydroxylation proceeds via a [3 + 2]-transition state to a cyclic ester intermediate which is

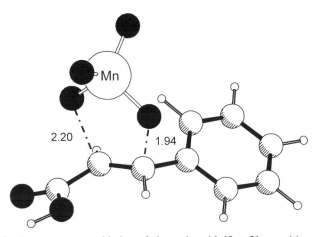

Fig. 8 Permanganate oxidation of cinnamic acid, [3 + 2]- transition state.

hydrolyzed in the course of the reaction. Freeman has carried out experiments on the permanganate ion oxidation[52,53] of nine unsaturated carboxylic acids **A–I** (Fig. 9) and the influence of substituents on the rate of the permanganate oxidation in phosphate buffered solutions (pH 6.83 ± 0.03). He has published free energy values derived from experimental kinetic data.[54]

Compounds **D** to **I** are all α, β unsaturated carboxylic acids with a changing number and position of alkyl substituents at the double bond. The chain length and position of the double bond is varied in compounds **A** to **C**. The rates of reaction were measured by monitoring spectral changes using a stopped flow spectrometer under pseudo-first-order conditions. The reactions were followed by monitoring the disappearance of the permanganate ion spectrally at 526, 584, or 660 nm and/or the rate of formation of colloidal manganese dioxide at 418 nm. The authors concluded from these measured reaction rates that the rate of oxidation is more sensitive to steric factors than to electronic effects. In regard to the reaction mechanism, it was suggested that both the [3 + 2]- and the [2 + 2]-transition state are in agreement with their experimental results.[54]

This systematic experimental study provided data for the calculation of the free energies of activation ($\Delta G^{\neq}_{exp}$) at standardized conditions from experimental rate constants against which the results obtained by calculation ($\Delta G^{\neq}_{calc}$) were compared. Two different basis sets have been employed in the DFT calculations: the split valence double-ζ (DZ) basis set 6-31G(d) with a triple-ζ (TZ)[55,56] valence basis set for manganese (this combination is named basis set I (**BS1**)) and the triple-ζ basis set 6-311 + G(d,p), which will be denoted basis set II (**BS2**). The results for transition states and intermediates on the BS1-level of theory are shown in Table 3, a graphical comparison of the free activation energies is shown in Fig. 10. *xyz*-Coordinates of all geometries are given in the supplementary material of Ref. 57.

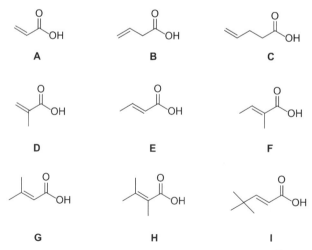

Fig. 9 Substrates for the permanganate oxidation.[54]

Table 3 Calculated free energy values $\Delta G^{\neq}$ for intermediates (GS) and transition states (TS) (**BS**1, in kcal/mol)[57]

Compound	$\Delta G_{exp}^{\neq}$	$\Delta G^{\neq}$ (3 + 2) TS	$\Delta G^{\neq}$ (2 + 2) TS	ΔG (3 + 2) GS	ΔG (2 + 2) GS
A	17.1	10.8	53.6	− 35.3	26.8
B	17.8	17.7	58.0	− 37.8	25.0
C	17.6	17.2	56.1	− 38.1	24.7
D	17.1	11.8	55.1	− 34.5	27.7
E	17.4	12.7	57.2	− 32.9	29.8
F	17.0	13.5	57.5	− 32.6	29.7
G	18.3	14.7	59.4	− 30.0	33.6
H	18.6	14.6	58.9	− 35.7	32.5
I	18.0	11.9	57.4	− 33.9	30.8

The geometries of the transition states differ significantly.[57] The transition state for the reaction of permanganate with alkene **A** shows an unsymmetrical geometry with calculated bond lengths of 2.29 and 1.90 Å (**A**$_{TS}$). For alkenes **B** and **C** the picture is quite contrary. Rather symmetrical transition states are calculated for **B**$_{TS}$ and **C**$_{TS}$ with bond lengths of 2.05/2.06 Å (**B**$_{TS}$) and 2.04/2.09 Å (**C**$_{TS}$) for the forming C–O bonds between the permanganate oxygens and the alkene carbons.

The symmetrical transition states **B**$_{TS}$ and **C**$_{TS}$ are very similar to the transition state calculated for the permanganate oxidation of ethene,[47] indicating that the substitutent does not play a major role. All other transition states are very

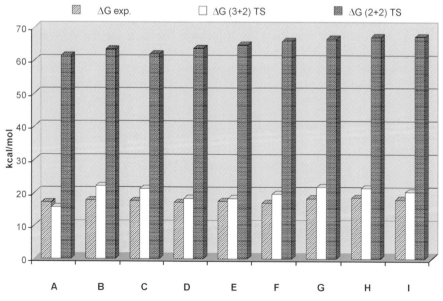

Fig. 10 Free activation energies for substrates **A**–**I**.

unsymmetrical, but concerted. The geometries of the transition states of D_{TS}–I_{TS} are similar regardless of the number of substituents. The bond lengths of the C–O forming bonds are independent from the substituents, whether it is a methyl group in α-position (E_{TS}, 2.30/1.91 Å), β-position (D_{TS}, 2.31/1.92 Å) or a *tert.*-butyl group (I_{TS}, 2.29/1.93 Å). Even three methyl groups as in H_{TS} do not distort the geometry more than what was observed for A_{TS}.

The higher level of theory used in the form of the **BS2** basis set does not result in major changes of the calculated geometries, but improves the agreement between experimental data and computational results (Table 4). Deviations in the bond length of 0.01 Å as well as changes of torsional angles of up to 4° are observed.

We also investigated the influence of solvation using the PCM model of Tomasi and co-workers, because certainly the interaction in solution is weaker, where the negative charge of the permanganate ion is going to be solvated, especially in aqueous solution. The activation energies for the solvated [3 + 2]-transition states are in reasonable agreement, although they are in general lower than the experimental values (with the exception of compound **H**). As can be seen from Table 5 the [3 + 2]-pathway is favored and, compared to the gas phase calculations, the deviations are much smaller.

Freeman chose substrates **D–I** to study the influence of steric bulk on the free energy of activation for the permanganate dihydroxylation. It is difficult to separate these substituent effects on the reaction rate, because it is not clear how to determine the relative contributions of steric and electronic effects to the overall observed effect. However, some comparisons are useful. For example, in the case of *trans*-crotonic acid **D** and 4,4-dimethyl-*trans*-2-pentenoic acid **I** the large bulk of the *t*-butyl group compared to that of a methyl group should result in a dominance of the steric over the electronic substituent effect.

It can be concluded that the [3 + 2]-pathway seems to be the only possible reaction pathway for the dihydroxylation by permanganate. There is a large difference in the activation enthalpies for these two pathways, and the intermediate of the [2 + 2]-addition of permanganate to ethylene is higher in energy than the

Table 4 Basis set dependence of calculated free energies (in kcal/mol)[57]

Compound	$\Delta G^{\neq}_{exp}$	$\Delta G^{\neq}$ (3 + 2) TS BS2	$\Delta G^{\neq}$ (3 + 2) TS BS1	$\Delta G^{\neq}$ (2 + 2) TS BS2	$\Delta G^{\neq}$ (2 + 2) TS BS1
A	17.1	15.7	10.8	61.5	53.6
B	17.8	22.0	17.7	63.4	58.0
C	17.6	21.3	17.2	62.0	56.1
D	17.1	18.3	11.8	63.7	55.1
E	17.4	18.3	12.7	64.8	57.2
F	17.0	19.6	13.5	65.9	57.5
G	18.3	21.8	14.7	66.7	59.4
H	18.6	21.4	14.6	67.2	58.9
I	18.0	20.3	11.9	67.3	57.4

Table 5 Free activation energies (**BS1**, PCM solvation model) for all transition states (in kcal/mol)[57]

Compound	A	B	C	D	E	F	G	H	I
$\Delta G^{\neq}_{exp}$	17.1	17.8	17.6	17.1	17.4	17.0	18.3	18.6	18.0
$\Delta G^{\neq}$ (3 + 2)TS PCM	15.4	15.0	16.3	16.2	17.3	15.5	17.4	20.6	17.0
$\Delta G^{\neq}$ (2 + 2)TS PCM	52.6	61.4	54.4	54.1	53.5	54.3	57.4	61.4	55.5

transition state for the corresponding [3 + 2]-addition reaction. A comparison of the calculated and experimentally determined free energies of activation for permanganate oxidation of α, β-unsaturated carboxylic acids shows that the energies calculated for [3 + 2]-mechanism are in better agreement with experimental values. Finally, experimentally determined KIEs for cinnamic acid are in good agreement with calculated isotope effects for the [3 + 2]-pathway and the calculated activation enthalpy is within the error limits of the experimental value.[51] It can therefore be concluded that a pathway via an oxetane intermediate is not viable.

Replacement of one oxygen by a chlorine changes the situation dramatically. Limberg has shown that the product of the addition of MnO_3Cl to ethylene is more stable in the triplet state and that the product distribution can be explained in terms of reaction channels.[58]

RUTHENIUM TETROXIDE (RUO$_4$)

The oxidation of alkenes by ruthenium tetroxide is another example of a metal-oxo oxidation reaction where the mechansim was discussed. This compound is known to dihydroxylate alkenes and several groups have investigated its reactivity.[59–67]

Theoretical work on this reaction was published by Norrby and co-workers in studies which focused mainly on the mechanism for the reaction of osmium tetroxide.[68] It was reported that the base complex $RuO_4 \cdot NH_3$ is 31 kJ/mol lower in energy compared to the isolated starting compounds, and that the formation of glycolates is largely favored over formation of oxetanes. The formation of the base-free oxetane is exothermic by 12 kJ/mol, while the formation of the glycol is exothermic by 231 kJ/mol. In the case of the oxetane the profile just shifts with the coordination of the base to − 40 kJ/mol, but is more exothermic in the glycol case with − 308 kJ/mol. Unfortunately neither the activation barriers nor free energies are given in the paper, but the similarity of the results for the intermediates and the well-known reactivity of the compounds lead to the conclusion that also in the case of ruthenium tetroxide the cis-dihydroxylation most probably proceeds via a [3 + 2]-pathway to the glycolate.

At present, only one system is known in which a the [2 + 2]-addition of metal tetroxides is preferred over the [3 + 2]-addition reaction. This is addition of

$(R_3PN)ReO_3$ to ketenes $R_2C{=}C{=}O$, for which a [2 + 2]-cycloaddition of a transition metal oxide across a C=C double bond has been reported.[69]

3 Epoxidation

Epoxides are a very versatile class of compounds and the interest in catalytic epoxidation reactions is very high.[70,71] They are the key raw materials in the syntheses of a wide variety of chemicals. A number of compounds have been shown to be catalytically active, but the regular laboratory reagents for epoxidations are generally methyl trioxorhenium(VII)[72–81] and the Jacobsen–Katsuki-catalysts[82–94] which can even introduce chirality. They are also theoretically well investigated[95–106] and are described below.

Chromylchloride (CrO_2Cl_2) has been included in this chapter for a different reason. It was the reagent which initiated an interesting discussion about the [3 + 2] vs. [2 + 2] reaction mechanism, a debate which continues for chromylchloride without resolution to this day. In addition to above mentioned compounds quite recently structurally related Mo(VI) and W(VI) peroxo complexes have been the subject of experimental and theoretical investigations.[97,107,108]

Despite extensive studies the detailed mechanism of the oxygen transfer remains controversial. There is general agreement that the high selectivities for a wide range of substrates and the stereospecifity of the reaction requires it to be heterolytic in nature. The rate-determining step has to be the oxygen transfer step from an alkylperoxometal complex to the double bond. Several pathways have been proposed since the original mechanistic proposal (Scheme 5) by Sheldon in 1973.[109,110]

On the basis of steric arguments this proposal was modified and it was suggested that the alkylperoxo ligand should coordinate through the distal rather then the proximal oxygen (Scheme 6). The exceptional reactivities can more easily be explained if the oxygen transfer proceeds from an alkylperoxometal group to the double bond of an allylic alcohol ligand.[111]

An alternative mechanism (Scheme 7) was proposed by Mimoun, which involves olefin coordination to the metal.[112–116] The insertion of the alkene into the metal–oxygen bond is thought to be the rate-limiting step of this reaction mechanism.

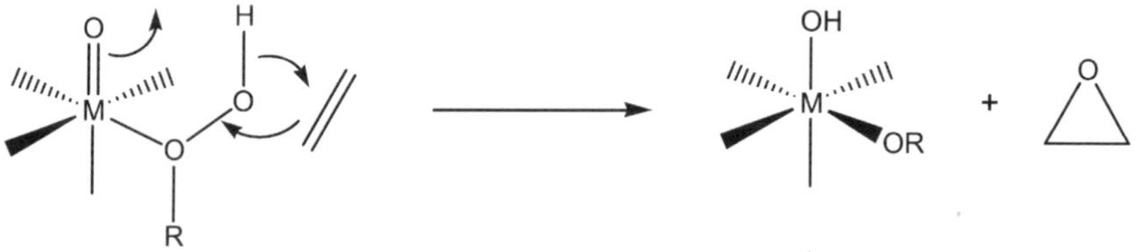

Scheme 5 Sheldon epoxidation mechanism.[109,110]

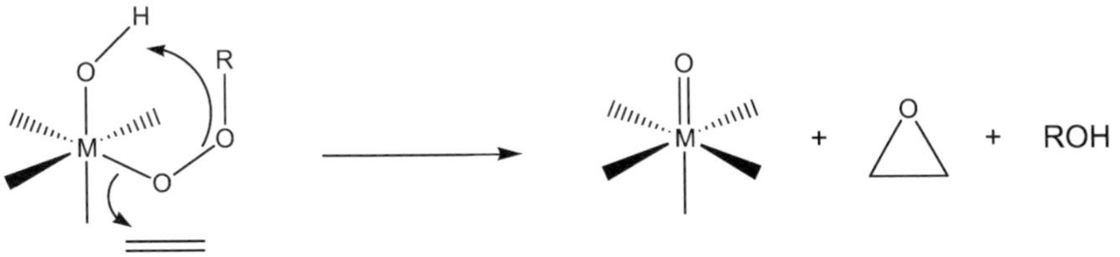

Scheme 6   Modified epoxidation mechanism.[111]

The current understanding for molybdenum and rhenium is that the involved species are bisperoxo- and hydroperoxo complexes, and that direct nucleophilic attack of the olefin at an electrophilic peroxo oxygen is significantly preferred over the two-step insertion mechanism proposed by Mimoun.

MOLYBDENUM

Several groups have shown that molybdenum(VI) peroxo complexes are active catalysts of alkene epoxidation reactions.[107,117–124] Metal activated hydroperoxo complexes were found to be the active species for the catalytic systems H_3NMoO_3/H_2O_2 and H_3NOMoO_3/H_2O_2 by hybrid DFT calculations. The hydroperoxo derivative with an additional axial Lewis-base ligand is about 9 kcal/mol more stable than the corresponding bisperoxo complex.[97] The activation barriers for the formation of the most stable peroxo and hydroperoxo intermediates are significantly higher than for the corresponding rhenium systems.

RHENIUM

Several complexes of rhenium(VII) with alkenes have also been studied in detail. Examples include compounds of the stoichiometry $LReO_3$, studied by Rappe[125] (L = Cp, Cp*, CH_3, OH,...) and Frenking[126] as well as by Rösch (L = CH_3).[30,96] Rappe investigated these compounds in order to determine whether dihydroxylation or epoxidation is the preferred pathway while Rösch studied the MTO epoxidation pathways in great detail.[96,99,100]

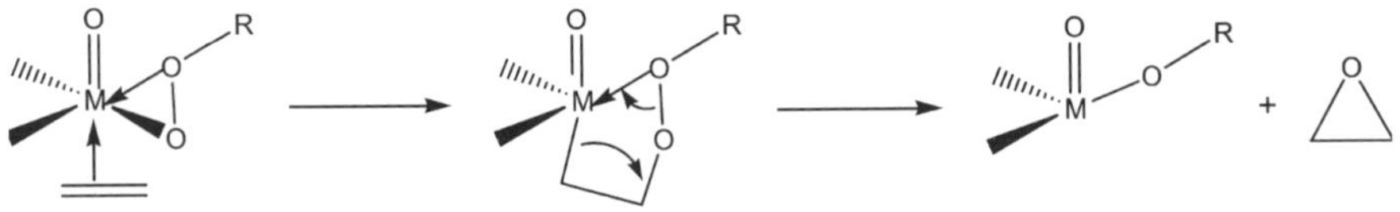

Scheme 7   Two-step mechanism proposed by Mimoun.[112–116]

Methyltrioxorhenium (CH₃)ReO₃

Methyl trioxorhenium (MTO) has proven to be an efficient catalyst in the presence of hydrogen peroxide, which leads to the formation of mono- and bisperoxo compounds; an additional aquo ligand has been found to stabilize the latter complex. MTO as well as the corresponding monoperoxo- and bisperoxo complexes have been studied, in both their free and monohydrated forms.[96]

As expected, hydration leads to a significant stabilization of the rhenium metal center. The calculated energy contribution is similar in the case of the monoperoxo compound and of MTO itself, but it is almost twice as high as for the bisperoxo complex, which underlines the stability of this complex. The calculations were also able to reproduce the geometry of an available X-ray geometry of the bisperoxo complex.[127] The structures of the transition states for direct oxygen transfer to the alkene and the insertion pathway via a [2 + 2]-like arrangement have been calculated as well as for the reaction of a hydroperoxy derivative.

The calculated barriers clearly provide better evidence for the peroxo pathway than for the Mimoun mechanism. The energies of the transition states for the insertion reaction are more than 12 kcal/mol higher than for the monoperoxo and bisperoxo complexes. The proposed precoordination of the alkene to the metal[112] could not be found. All hydroperoxo transition states are rather high in energy, with barriers of 30 kcal/mol and higher. For the various peroxo complexes (with and

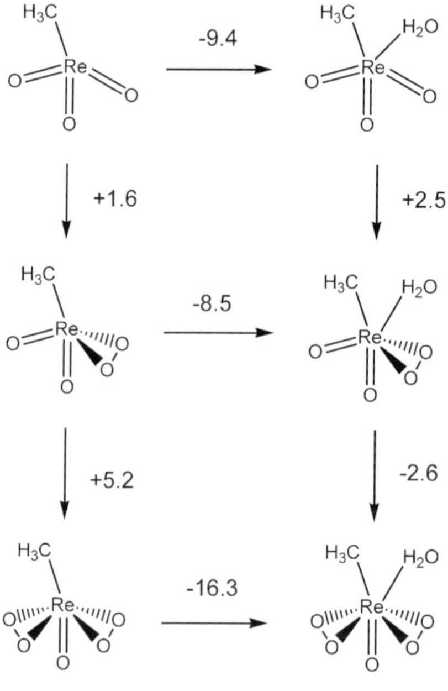

Fig. 11 MTO and its monoperoxo and bisperoxo complexes as well as monohydrated complexes. Energies given in kcal/mol.[96]

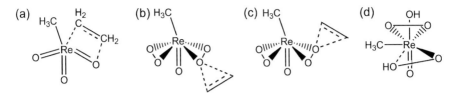

Fig. 12 Transition states for the insertion pathway (a), direct oxygen transfer (b,c) and hydroperoxo species (d).

without the water ligand) activation barriers of 12.4 kcal/mol ($H_3C-Re(O)(O_2)_2$), 16.2 ($H_3C-Re(O)_2(O_2)\cdot H_2O$) and higher have been calculated. Depending on the reaction conditions the epoxidation of alkenes catalyzed by the system MTO/H_2O_2 is found to proceed by direct oxygen transfer via one of the peroxo transition states.

In the case of epoxidation of allylic alcohols, a recent theoretical study concluded that hydrogen-bonding plays an important role in stabilizing the transition state, and that the barriers for the formation of solvated bisperoxo complexes are lower than for formation of the corresponding monoperoxo complexes. This is in agreement with experimental results.[95]

JACOBSEN–KATSUKI-CATALYSTS

The Jacobsen–Katsuki-catalysts (Fig. 13) have recently received much attention as the most widely used alkene epoxidation catalysts. An example of Jacobsen's manganese-salen catalyst is shown in Fig. 13. They promote the stereoselective conversion of prochiral olefins to chiral epoxides with enantiomeric excesses regularly better than 90% and sometimes exceeding 98%.[82,89,92,93,128] The oxidation state of the metal changes during the catalytic cycle as shown in Scheme 8.

The *cis–trans* alkene isomerization observed in the transformation of conjugated alkenes gave rise to proposals of several different reaction mechanisms. Four different pathways for the epoxidation of alkenes by the manganese metal-oxo species have been proposed (Scheme 9) in order to explain the experimental results. The two concerted pathways (b,c) involve formation of a three-membered or four-membered transition state that involves the alkene and either the transferring oxygen or the Mn=O, respectively, of the formal Mn(V) species. The other pathways are stepwise mechanisms; one proceeds via bond formation between the alkene and the

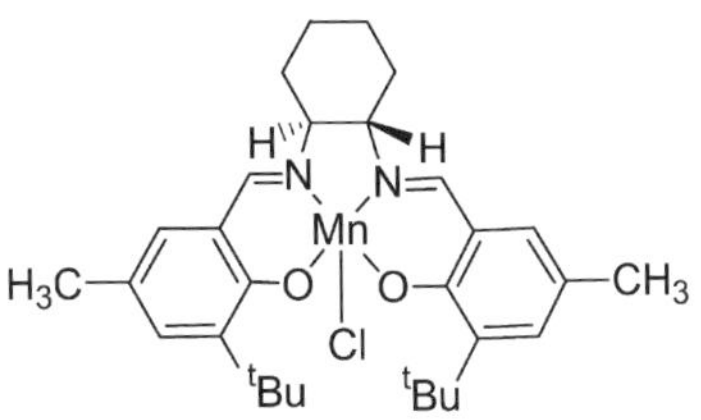

Fig. 13 Jacobsen–Katsuki-catalyst with a cyclohexane backbone.

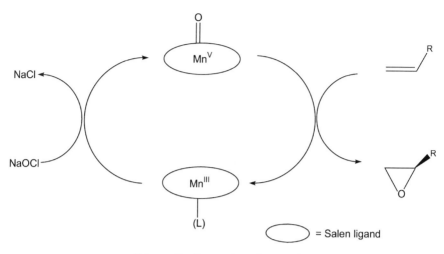

Scheme 8 Proposed catalytic cycle.

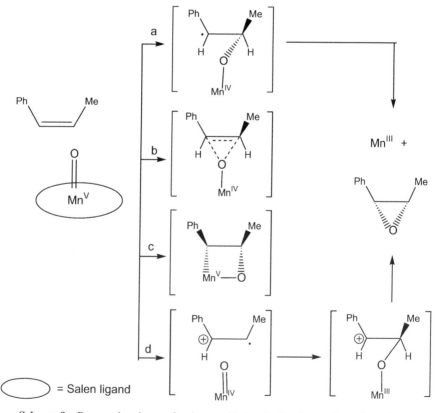

Scheme 9 Proposed pathways for the epoxidation by Jacobsen–Katsuki-catalysts.

oxygen to form a Mn(IV) diradical intermediate (a), while the other (d) involves an initial single electron transfer to create an alkene radical cation and a Mn(IV) species followed by bond formation to oxygen to give a Mn(III) cation intermediate.[84]

The importance of steric bulk in the 3,3'-positions of the salen ligand and the influence of electronic effects of substituents in the 5,5'-position were recognized at an early stage.[86] Three major pathways for the approach of the olefin have been proposed and are shown in Fig. 14. Jacobsen originally proposed[92] approach **b** to account for the effects of substituents R_2 and to avoid steric interactions between alkenes and bulky *t*-butyl groups at R_3. He then modified this to pathway **a** when it was found that a dimethylcyclohexyl group at R_1–R_1 gave a lower stereoselectivity than a cyclohexyl group.[82] Pathway **c** has been predicted on the basis of force-field modeling by Kasuga.[129]

The current understanding is that the multiplicity of the intermediate influences whether a stepwise or concerted reaction can occur, although spin–orbit coupling will accelerate intersystem crossing during reactions involving multiplicity changes. The manganese(III) systems are high-spin quintet ground states, while the active Mn(V)-oxo species may be either in the quintet or triplet state, depending on the nature of the ligand L.[104] The calculated model systems Mn^{III}(ligand), Mn^{III}(ligand)Cl, Mn^V(ligand), Mn^V(ligand)Cl are shown in Fig. 15.

The manganese systems can have different electronic configurations: singlet (*s*), triplet (*t*) and quintet (*q*). DFT calculations were able to provide geometries which are in very good agreement with available experimental X-ray structures, taking into account the differences between the conjugated catalyst und our model system.

All different spin states have been calculated in the ground state and the calculations indicate that the potential structure of the active Jacobsen catalyst is the Mn^V(O)(ligand)Cl_*t* complex shown in Fig. 17.[104]

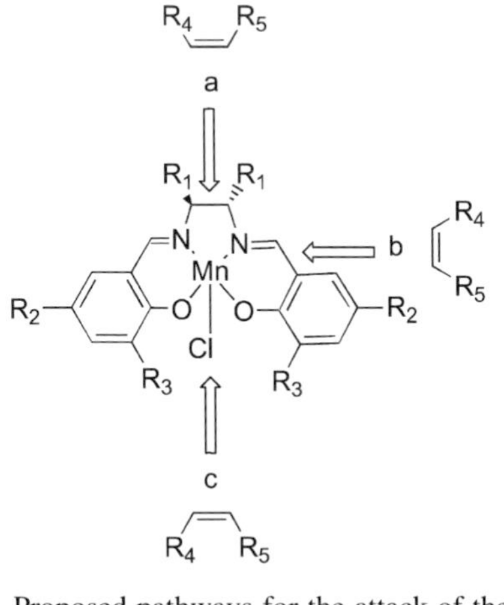

Fig. 14 Proposed pathways for the attack of the alkene.

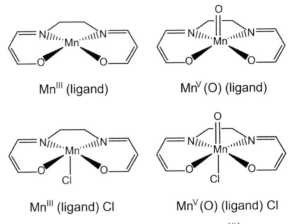

Fig. 15 Calculated model systems.[104]

The importance of the counter-ion has been recognized. The counter-ion influences the spin multiplicities and, potentially, the relative rates of the stereospecific concerted and stepwise non-concerted processes. The calculations indicate that the active manganese-oxo species in the Jacobsen epoxidation may either be a high-spin quintet or triplet, depending on the ligand.

The active species has also been observed experimentally by electrospray mass spectroscopy[106,130–133] and the importance of the position of the transition state (relative to reactant and product) on the stereoselectivity was discussed.[86]

The first report of calculated transition state structures raised the issue that there may be a change in the spin-state during the reaction.[134] The lowest energy spin-state for the starting material is a triplet, while the product is significantly more

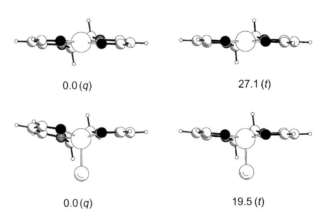

Fig. 16 Optimized geometries for the manganese(III) complexes in the quintet (*q*) and triplet (*t*) state, energies in kcal/mol.

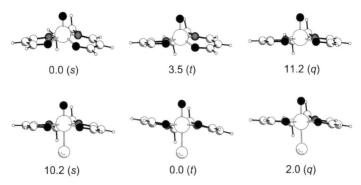

0.0 (s)   3.5 (t)   11.2 (q)

10.2 (s)   0.0 (t)   2.0 (q)

Fig. 17 Optimized geometries for the manganese(V) complexes in the quintet (q), triplet (t) and singulet (s) state, energies in kcal/mol.

stable in the quintet state. A reaction that proceeds exclusively on the quintet energy surface, and a reaction that leads to formation of an oxetane intermediate can both be ruled out because of their high activation barriers.

The electronic effect of substituents, ligands or substrates can lead to a different triplet–quintet (t–q) spin crossing which will affect the diastereoselectivity of the epoxidation reaction. This can be shown by a comparison of the effects of 1,2-dimethoxy and 1,2-dinitro substituents on the reactivity of ethene. The t–q energy difference decreases from 25 kcal/mol for $O_2N-C_2H_2-NO_2$ to 9.5 kcal/mol (C_2H_4) to only 3 kcal/mol for $H_3CO-C_2H_2-OCH_3$. The change in the spin-state that occurs on proceeding from the triplet reactant to the quintet product should occur later on the reaction coordinate for electron poor alkenes and allow for a bond rotation prior to formation of the epoxide product and therefore demonstrate a low diastereoselectivity. On the other hand, a higher diastereoselectivity is favored for electron-donating alkene substituents, because now the change in spin state will occur earlier along the reaction coordinate, before formation of radical intermediates.

The title "Radical intermediates in the Jacobsen–Katsuki Epoxidation" indicates that other authors predict this reaction to proceed via open-shell systems.[135] There is agreement with the previously mentioned study that the triplet state for the starting material is about 10 kcal/mol lower in energy compared to the quintet state and that a metallaoxetane is not a feasible intermediate. There is no agreement on the energy differences of the product: triplet and quintet are calculated to have approximately the same energy. These authors also calculated the structures of the transition state with the full salen ligand, but without significant changes of the energy differences to the model system. The study concludes that the reaction proceeds on the triplet surface.

The two-state-reactivity[136,137] with changes between spin states provide the best explanation for the differences in these experimental results, but there are still open questions to solve. For example, the significance of the observed oxo-bridged dimer,[106] and its role in the catalytic cycle are not understood.

CHROMIUM

A recent paper on the epoxidation of alkenes by a chromium(V) complex also reported a two-state reactivity.[138] The complex cp*Cr(O)Cl$_2$, shown in Fig. 18, epoxidized alkenes stoichiometrically. The reactant has one unpaired electron, corresponding to a doublet state and is converted into a chromium(III) product. The quartet state of the product is significantly more stable than the doublet state, indicating that the reaction hyper-surface changes from the doublet to the quartet state during the reaction. This reaction shows that chromium oxo species generated by the direct activation of oxygen are capable of epoxidizing olefins. Depending on the reaction conditions, chromylchloride, (CrO$_2$Cl$_2$) is also capable of oxidizing alkenes.

Chromylchloride CrO$_2$Cl$_2$

The oxidation of olefins by chromylchloride has been known since the 19th century. Even in the absence of peroxides, this reaction yields epoxides rather than diols in a complex mixture of products, which also contains *cis*-chlorohydrins and vicinal dichlorides. Many different reaction mechanisms have been proposed to explain the great variety of observed products, but none of the proposed intermediates have been identified. Stairs favors a direct interaction of the alkene with one oxygen atom of chromylchloride,[139–141] while Sharpless proposed a chromaoxetane[10] that forms via a [2 + 2]-pathway, a proposal which has led to intense discussions.

The results of early calculations using the general valence bond approach by Rappe[142–144] supported the conclusion that the chromaoxetane is a likely intermediate. This also contributed to the discussion of metallaoxetane intermediates derived from a [2 + 2]-cyclo addition which has been raised in almost every study of oxygen-transfer reactions ever since 1977.

Concise theoretical studies of Ziegler[145,146] analyzed all of the possible reaction pathways including the crossover from the singlet to the triplet surface with the transition state on the singlet surface while the formed product is a triplet species. It could be shown that the epoxide precursor is formed via a [3 + 2]-addition of ethylene to two Cr=O bonds followed by rearrangement to the epoxide product (Scheme 10).

The activation barriers for these two steps have been computed as +14.0 and +21.9 kcal/mol. The alternative pathways, direct addition of an oxygen to the ethylene and the [2 + 2]-addition both have a higher activation energy of +30.0 and

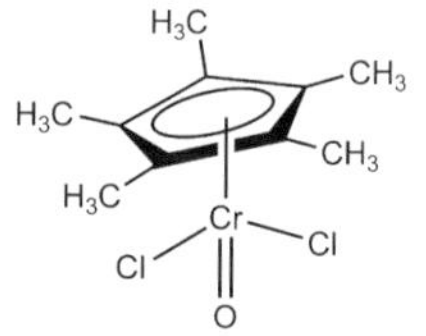

Fig. 18   Chromium(V)-oxo complex capable of epoxidizing alkenes stoichiometrically.[138]

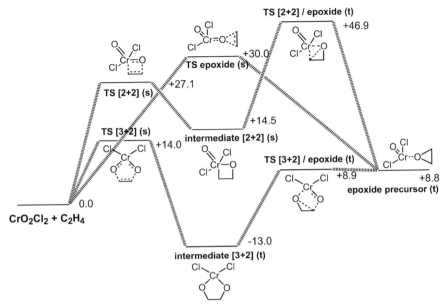

Scheme 10 Ziegler's results on the epoxide formation.[145]

+27.1 kcal/mol. The formation of the chlorohydrin products was also explained by a [3 + 2]-addition to one Cr–Cl and one Cr=O bond.[145]

The transition state with the lowest barrier leads to the [3 + 2]-intermediate, and this should result in the formation of diols, at least as side products. There are still open questions to address, and it is expected that research in this area will continue.

4 Summary

Theoretical investigations of transition metal mediated processes have improved our understanding of the mechanisms of a number of important reactions. These reactions often proceed via very reactive intermediates which are experimentally difficult to observe directly. DFT calculations have shown to be an extremely useful tool in the area of organometallic chemistry and especially for mechanistic investigations. Quantum chemical calculations can provide data on the electronic structure, energetics and the reaction intermediates. Using the data it is possible to distinguish between different proposed reaction mechanisms as could be shown above for the dihydroxylation reaction of osmium tetroxide. Computational chemistry could provide insight into many reaction pathways which could not be determined experimentally before. Many aspects still remain unsolved or have just begun to be explored. The future teamwork of experimental and theoretical chemists promises interesting results.

References

1. Wiberg, K.B. (1965). *Oxidation in Organic Chemistry*, Part A, vol. 5. Academic Press, New York
2. Adam, W., Saha-Moeller, C.R. and Ganeshpure, P.A. (2001). *Chem. Rev.* **101**, 3499–3548
3. Ahlrichs, R., Elliott, S.D. and Huniar, U. (1998). *Ber. Bunsen-Ges.* **102**, 795–804
4. Chermette, H. (1998). *Coord. Chem. Rev.* **178–180**, 699–721
5. Frenking, G., Antes, I., Boehme, C., Dapprich, S., Ehlers, A.W., Jonas, V., Neuhaus, A., Otto, M., Stegmann, R., Veldkamp, A. and Vyboishchikov, S.F. (1996). In *Reviews in Computational Chemistry*, Lipkowitz, K.B. and Boyd, D.B. (eds), vol. 8, pp. 63–144. Wiley VCH, New York
6. Jensen, F. (1999). *Introduction to Computational Chemistry*. Wiley, New York
7. Laird, B.B., Ross, R.B. and Ziegler, T. (1996). *ACS Symp. Ser.* **629**, 1–17
8. Torrent, M., Sola, M. and Frenking, G. (2000). *Chem. Rev.* **100**, 439–493
9. Diedenhofen, M., Wagener, T. and Frenking, G. (2001). In *Computational Organometallic Chemistry*, Cundari, T.R. (ed.), Marcel Dekker, New York
10. Sharpless, K.B., Teranishi, A.Y. and Bäckvall, J.E. (1977). *J. Am. Chem. Soc.* **99**, 3120–3128
11. Corey, E.J. and Lotto, G.I. (1990). *Tetrahedron Lett.* **31**, 2665–2668
12. Corey, E.J. and Noe, M.C. (1993). *J. Am. Chem. Soc.* **115**, 12579–12580
13. Corey, E.J., Noe, M.C. and Sarshar, S. (1993). *J. Am. Chem. Soc.* **115**, 3828–3829
14. Corey, E.J., Noe, M.C. and Grogan, M.J. (1994). *Tetrahedron Lett.* **35**, 6427–6430
15. Corey, E.J., Noe, M.C. and Lin, S. (1995). *Tetrahedron Lett.* **36**, 8741–8744
16. Corey, E.J. and Noe, M.C. (1996). *J. Am. Chem. Soc.* **118**, 11038–11053
17. Corey, E.J., Sarshar, S., Azimioara, M.D., Newbold, R.C. and Noe, M.C. (1996). *J. Am. Chem. Soc.* **118**, 7851–7852
18. Corey, E.J., Noe, M.C. and Grogan, M.J. (1996). *Tetrahedron Lett.* **37**, 4899–4902
19. Becker, H., Ho, P.T., Kolb, H.C., Loren, S., Norrby, P.-O. and Sharpless, K.B. (1994). *Tetrahedron Lett.* **35**, 7315–7318
20. Kolb, H.C., Andersson, P.G. and Sharpless, K.B. (1994). *J. Am. Chem. Soc.* **116**, 1278–1291
21. Kolb, H.C. and Sharpless, K.B. (1998). *Transition Met. Org. Synth.* **2**, 219–242
22. Nelson, D.W., Gypser, A., Ho, P.T., Kolb, H.C., Kondo, T., Kwong, H.-L., McGrath, D.V., Rubin, A.E., Norrby, P.-O., Gable, K.P. and Sharpless, K.B. (1997). *J. Am. Chem. Soc.* **119**, 1840–1858
23. Norrby, P.-O., Becker, H. and Sharpless, K.B. (1996). *J. Am. Chem. Soc.* **118**, 35–42
24. Vanhessche, K.P.M. and Sharpless, K.B. (1996). *J. Org. Chem.* **61**, 7978–7979
25. Göbel, T. and Sharpless, K.B. (1993). *Angew. Chem. Int. Ed. Engl.* **32**, 1329–1331
26. Pidun, U., Boehme, C. and Frenking, G. (1997). *Angew. Chem. Int. Ed. Engl.* **35**, 2817–2820
27. Dapprich, S., Ujaque, G., Maseras, F., Lledos, A., Musaev, D.G. and Morokuma, K. (1996). *J. Am. Chem. Soc.* **118**, 11660–11661
28. Torrent, M., Deng, L., Duran, M., Sola, M. and Ziegler, T. (1997). *Organometallics* **16**, 13–19
29. DelMonte, A.J., Haller, J., Houk, K.N., Sharpless, K.B., Singleton, D.A., Strassner, T. and Thomas, A.A. (1997). *J. Am. Chem. Soc.* **119**, 9907–9908
30. Gisdakis, P. and Rösch, N. (2001). *J. Am. Chem. Soc.* **123**, 697–701
31. Dobler, C., Mehltretter, G. and Beller, M. (1999). *Angew. Chem. Int. Ed. Engl.* **38**, 3026–3028
32. Dobler, C., Mehltretter, G.M., Sundermeier, U. and Beller, M. (2001). *J. Organomet. Chem.* **621**, 70–76

33. Mehltretter, G.M., Dobler, C., Sundermeier, U. and Beller, M. (2000). *Tetrahedron Lett.* **41**, 8083–8087
34. Doebler, C., Mehltretter, G.M., Sundermeier, U. and Beller, M. (2000). *J. Am. Chem. Soc.* **122**, 10289–10297
35. Criegee, R. (1936). *Justus Liebigs Ann. Chem.* **522**, 75–96
36. Jorgensen, K.A. and Hoffmann, R. (1986). *J. Am. Chem. Soc.* **108**, 1867–76
37. Criegee, R., Marchand, B. and Wannowius, H. (1942). *Liebigs Ann. Chem.* **550**, 99–133
38. Singleton, D.A. and Thomas, A.A. (1995). *J. Am. Chem. Soc.* **117**, 9357–9358
39. Ujaque, G., Maseras, F. and Lledos, A. (1997). *J. Org. Chem.* **62**, 7892–7894
40. Norrby, P.-O., Rasmussen, T., Haller, J., Strassner, T. and Houk, K.N. (1999). *J. Am. Chem. Soc.* **121**, 10186–10192
41. Maseras, F. (1999). *Top. Organomet. Chem.* **4**, 165–191
42. Sharpless, K.B., Amberg, W., Beller, M., Chen, H., Hartung, J., Kawanami, Y., Lubben, D., Manoury, E., Ogino, Y., *et al.* (1991). *J. Org. Chem.* **56**, 4585–4588
43. Sharpless, K.B., Amberg, W., Bennani, Y.L., Crispino, G.A., Hartung, J., Jeong, K.S., Kwong, H.L., Morikawa, K., Wang, Z.M., *et al.* (1992). *J. Org. Chem.* **57**, 2768–2771
44. Sheldon, R.A., Kochi, J.K. (1981). *Metal-Catalyzed Oxidations of Organic Compounds* Academic Press, New York
45. Waters, W.A. (1958). *Quart. Rev. (London)* **12**, 277–300
46. Henbest, H.B., Jackson, W.R. and Robb, B.C.G. (1966). *J. Chem. Soc. B Phys. Org.* 803–807
47. Houk, K.N. and Strassner, T. (1999). *J. Org. Chem.* **64**, 800–802
48. Lee, D.G. and Brownridge, J.R. (1974). *J. Am. Chem. Soc.* **96**, 5517–5523
49. Lee, D.G. and Brown, K.C. (1982). *J. Am. Chem. Soc.* **104**, 5076–5081
50. Lee, D.G., and Nagarajan, K. (1985). *Can. J. Chem.* **63**, 1018–1023
51. Strassner, T. (2003). In preparation
52. Simandi, L.I., Jaky, M., Freeman, F., Fuselier, C.O. and Karchefski, E.M. (1978). *Inorg. Chim. Acta* **31**, L457–L459
53. Freeman, F. (1973). *Rev. Reactive Species Chem. React.* **1**, 37–64
54. Freeman, F. and Kappos, J.C. (1986). *J. Org. Chem.* **51**, 1654–1657
55. Schaefer, A., Huber, C. and Ahlrichs, R. (1994). *J. Chem. Phys.* **100**, 5829–5835
56. Frenking, G., Antes, I., Boehme, M., Dapprich, S., Ehlers, A.W., Jonas, V., Neuhaus, A., Otto, M., Stegmann, R., *et al.* (1996). *Rev. Comput. Chem.* **8**, 63–144
57. Strassner, T. and Busold, M. (2001). *J. Org. Chem.* **66**, 672–676
58. Wistuba, T. and Limberg, C. (2001). *Chem.-Eur. J.* **7**, 4674–4685
59. Albarella, L., Piccialli, V., Smaldone, D. and Sica, D. (1996). *J. Chem. Res. Synopses* 400–401
60. Albarella, L., Giordano, F., Lasalvia, M., Piccialli, V. and Sica, D. (1995). *Tetrahedron Lett.* **36**, 5267–5270
61. Shing, T.K.M., Tai, V.W.F. and Tam, E.K.W. (1994). *Angew. Chem.* **106**, 2408–2409
62. El-Hendawy, A.M., Griffith, W.P., Taha, F.I. and Moussa, M.N. (1989). *J. Chem. Soc. Dalton Trans.* 901–906
63. Ayres, D.C. and Levy, D.P. (1986). *Tetrahedron* **42**, 4259–4265
64. Carlsen, P.H.J., Katsuki, T., Martin, V.S. and Sharpless, K.B. (1981). *J. Org. Chem.* **46**, 3936–3938
65. Foglia, T.A., Barr, P.A. and Malloy, A.J. (1977). *J. Am. Oil Chem. Soc.* **54**, 858A–861A
66. Sharpless, K.B. and Akashi, K. (1976). *J. Am. Chem. Soc.* **98**, 1986–1987
67. Rylander, P.N. (1969). *Engelhard Ind. Tech. Bull.* **9**, 135–138
68. Norrby, P.O., Kolb, H.C. and Sharpless, K.B. (1994). *Organometallics* **13**, 344–347
69. Deubel, D.V. and Frenking, G. (1999). *J. Am. Chem. Soc.* **121**, 2021–2031

70. Yudanov, I.V., Di Valentin, C., Gisdakis, P. and Rösch, N. (2000). *J. Mol. Catal. A: Chem.* **158**, 189–197
71. Valentin, C.D., Gisdakis, P., Yudanov, I.V. and Rösch, N. (2000). *J. Org. Chem.* **65**, 2996–3004
72. Al-Ajlouni, A.M. and Espenson, J.H. (1996). *J. Org. Chem.* **61**, 3969–3976
73. Zhu, Z. and Espenson, J.H. (1995). *J. Mol. Catal. A: Chem.* **103**, 87–94
74. Herrmann, W.A., Kiprof, P., Rypdal, K., Tremmel, J., Blom, R., Alberto, R., Behm, J., Albach, R.W., Bock, H., *et al.* (1991). *J. Am. Chem. Soc.* **113**, 6527–6537
75. Herrmann, W.A., Fischer, R.W. and Marz, D.W. (1991). *Angew. Chem.* **103**, 1706–1709
76. Herrmann, W.A. (1995). *J. Organomet. Chem.* **500**, 149–174
77. Herrmann, W.A., Kühn, F.E., Mattner, M.R., Artus, G.R.J., Geisberger, M.R. and Correia, J.D.G. (1997). *J. Organomet. Chem.* **538**, 203–209
78. Herrmann, W.A. and Kühn, F.E. (1997). *Acc. Chem. Res.* **30**, 169–180
79. Herrmann, W.A., Correia, J.D.G., Rauch, M.U., Artus, G.R.J. and Kühn, F.E. (1997). *J. Mol. Catal. A: Chem.* **118**, 33–45
80. Kühn, F.E. and Herrmann, W.A. (2000). *Struct. Bond (Berlin)* **97**, 213–236
81. Wikrent, P., Drouin, B.J., Kukolich, S.G., Lilly, J.C., Ashby, M.T., Herrmann, W.A. and Scherer, W. (1997). *J. Chem. Phys.* **107**, 2187–2192
82. Jacobsen, E.N., Zhang, W., Muci, A.R., Ecker, J.R. and Deng, L. (1991). *J. Am. Chem. Soc.* **113**, 7063–7064
83. Lee, N.H. and Jacobsen, E.N. (1991). *Tetrahedron Lett.* **32**, 6533–6536
84. Linker, T. (1997). *Angew. Chem. Int. Ed. Engl.* **36**, 2060–2062
85. Palucki, M., McCormick, G.J. and Jacobsen, E.N. (1995). *Tetrahedron Lett.* **36**, 5457–5460
86. Palucki, M., Finney, N.S., Pospisil, P.J., Gueler, M.L., Ishida, T. and Jacobsen, E.N. (1998). *J. Am. Chem. Soc.* **120**, 948–954
87. Pospisil, P.J., Carsten, D.H. and Jacobsen, E.N. (1996). *Chem.-Eur. J.* **2**, 974–980
88. Senanayake, C.H. and Jacobsen, E.N. (1999). *Process Chem. Pharm. Ind.* 347–368
89. Jacobsen, E.N., Zhang, W. and Guler, M.L. (1991). *J. Am. Chem. Soc.* **113**, 6703–6704
90. Jacobsen, E.N. (1993). *Catal. Asymm. Synth.* 159–202
91. Jacobsen, E.N., Deng, L., Furukawa, Y. and Martinez, L.E. (1994). *Tetrahedron* **50**, 4323–4334
92. Zhang, W., Loebach, J.L., Wilson, S.R. and Jacobsen, E.N. (1990). *J. Am. Chem. Soc.* **112**, 2801–2803
93. Zhang, W. and Jacobsen, E.N. (1991). *J. Org. Chem.* **56**, 2296–2298
94. Jacobsen, E.N. and Wu, M.H. (1999). *Compr. Asymmetric Catal. I–III* **2**, 649–677
95. Di Valentin, C., Gandolfi, R., Gisdakis, P. and Rösch, N. (2001). *J. Am. Chem. Soc.* **123**, 2365–2376
96. Gisdakis, P., Antonczak, S., Köstlmeier, S., Herrmann, W.A. and Rösch, N. (1998). *Angew. Chem. Int. Ed. Engl.* **37**, 2211–2214
97. Gisdakis, P., Yudanov, I.V. and Rösch, N. (2001). *Inorg. Chem.* **40**, 3755–3765
98. Gisdakis, P., and Rösch, N. (2001). *Eur. J. Org. Chem.* 719–723
99. Köstlmeier, S., Haeberlen, O.D., Rösch, N., Herrmann, W.A., Solouki, B. and Bock, H. (1996). *Organometallics* **15**, 1872–1878
100. Köstlmeier, S., Nasluzov, V.A., Herrmann, W.A. and Rösch, N. (1997). *Organometallics* **16**, 1786–1792
101. Kühn, F.E., Santos, A.M., Roesky, P.W., Herdtweck, E., Scherer, W., Gisdakis, P., Yudanov, I.V., Di Valentin, C. and Rösch, N. (1999). *Chem.-Eur. J.* **5**, 3603–3615
102. Rösch, N., Gisdakis, P., Yudanov, I.V. and Di Valentin, C. (2000). *Peroxide Chem.* 601–619
103. Jacobsen, H. and Cavallo, L. (2001). *Chem.-Eur. J.* **7**, 800–807

104. Strassner, T. and Houk, K.N. (1999). *Org. Lett.* **1**, 419–421
105. El-Bahraoui, J., Wiest, O., Feichtinger, D. and Plattner, D.A. (2001). *Angew. Chem. Int. Ed. Engl.* **40**, 2073–2076
106. Plattner, D.A., Feichtinger, D., El-Bahraoui, J. and Wiest, O. (2000). *Int. J. Mass Spectr.* **195**, 351–362
107. Kühn, F.E., Groarke, M., Bencze, E., Herdtweck, E., Prazeres, A., Santos, A.M., Calhorda, M.J., Romao, C.C., Goncalves, I.S., Lopes, A.D. and Pillinger, M. (2002). *Chem.-Eur. J.* **8**, 2370–2383
108. Di Valentin, C., Gisdakis, P., Yudanov, I.V. and Rösch, N. (2000). *J. Org. Chem.* **65**, 2996–3004
109. Sheldon, R.A. (1973). *Recl. Trav. Chim. Pays-Bas* **92**, 253–266
110. Sheldon, R.A. (1973). *Recl. Trav. Chim. Pays-Bas* **92**, 367–373
111. Chong, A.O. and Sharpless, K.B. (1977). *J. Org. Chem.* **42**, 1587–1590
112. Mimoun, H., Seree de Roch, I. and Sajus, L. (1970). *Tetrahedron* **26**, 37–50
113. Mimoun, H. (1987). *Catal. Today* **1**, 281–295
114. Mimoun, H., Mignard, M., Brechot, P. and Saussine, L. (1986). *J. Am. Chem. Soc.* **108**, 3711–3718
115. Mimoun, H. (1982). *Angew. Chem.* **94**, 750–766
116. Mimoun, H. (1980). *J. Mol. Catal.* **7**, 1–29
117. Thiel, W.R. and Eppinger, J. (1997). *Chem.-Eur. J.* **3**, 696–705
118. Thiel, W.R. (1997). *J. Mol. Catal. A: Chem.* **117**, 449–454
119. Thiel, W.R. and Priermeier, T. (1995). *Angew. Chem. Int. Ed. Engl.* **34**, 1737–1738
120. Wahl, G., Kleinhenz, D., Schorm, A., Sundermeyer, J., Stowasser, R., Rummey, C., Bringmann, G., Fickert, C. and Kiefer, W. (1999). *Chem.-Eur. J.* **5**, 3237–3251
121. Thiel, W.R., Angstl, M. and Priermeier, T. (1994). *Chem. Ber.* **127**, 2373–2379
122. Thiel, W.R., Angstl, M. and Hansen, N. (1995). *J. Mol. Catal. A: Chem.* **103**, 5–10
123. Thiel, W.R. (2002). *Inorg. Chem. Highlights* 123–138
124. Hroch, A., Gemmecker, G. and Thiel, W.R. (2000). *Eur. J. Inorg. Chem.* 1107–1114
125. Pietsch, M.A., Russo, T.V., Murphy, R.B., Martin, R.L. and Rappe, A.K. (1998). *Organometallics* **17**, 2716–2719
126. Deubel, D.V., Schlecht, S. and Frenking, G. (2001). *J. Am. Chem. Soc.* **123**, 10085–10094
127. Herrmann, W.A., Fischer, R.W., Scherer, W. and Rauch, M.U. (1993). *Angew. Chem.* **105**, 1209–1212
128. Zhang, W., Lee, N.H. and Jacobsen, E.N. (1994). *J. Am. Chem. Soc.* **116**, 425–426
129. Houk, K.N., DeMello, N.C., Condroski, K., Fennen, J. and Kasuga, T. (1997). *Electron. Conf. Heterocycl. Chem. (Proc.)*
130. Plattner, D.A. (2001). *Int. J. Mass Spectr.* **207**, 125–144
131. Feichtinger, D. and Plattner, D.A. (2001). *Chem.-Eur. J.* **7**, 591–599
132. Feichtinger, D. and Plattner, D.A. (2000). *Perkin 2*, 1023–1028
133. Feichtinger, D. and Plattner, D.A. (1997). *Angew. Chem. Int. Ed. Engl.* **36**, 1718–1719
134. Linde, C., Aakermark, B., Norrby, P.-O. and Svensson, M. (1999). *J. Am. Chem. Soc.* **121**, 5083–5084
135. Cavallo, L. and Jacobsen, H. (2000). *Angew. Chem. Int. Ed. Engl.* **39**, 589–592
136. Schröder, D., Shaik, S. and Schwarz, H. (2000). *Acc. Chem. Res.* **33**, 139–145
137. Shaik, S., Danovich, D., Fiedler, A., Schröder, D. and Schwarz, H. (1995). *Helv. Chim. Acta* **78**, 1393–1407
138. Hess, J.S., Leelasubcharoen, S., Rheingold, A.L., Doren, D.J. and Theopold, K.H. (2002). *J. Am. Chem. Soc.* **124**, 2454–2455
139. Makhija, R.C., and Stairs, R.A. (1968). *Can. J. Chem.* **46**, 1255–1260
140. Makhija, R.C., and Stairs, R.A. (1969). *Can. J. Chem.* **47**, 2293–2299

141. Stairs, R.A., Diaper, D.G., and Gatzke, A.L. (1963). *Can. J. Chem.* **41**, 1059
142. Rappe, A.K. and Goddard, W.A., III. (1980). *Nature (London)* **285**, 311–312
143. Rappe, A.K. and Goddard, W.A., III. (1980). *J. Am. Chem. Soc.* **102**, 5114–5115
144. Rappe, A.K. and Goddard, W.A., III. (1982). *J. Am. Chem. Soc.* **104**, 3287–3294
145. Torrent, M., Deng, L., Duran, M., Sola, M., and Ziegler, T. (1999). *Can. J. Chem.* **77**, 1476–1491
146. Torrent, M., Deng, L. and Ziegler, T. (1998). *Inorg. Chem.* **37**, 1307–1314

Solvent effects, reaction coordinates, and reorganization energies on nucleophilic substitution reactions in aqueous solution

JIALI GAO, MIREIA GARCIA-VILOCA, TINA D. POULSEN and YIRONG MO[*]

Department of Chemistry and Supercomputing Institute, University of Minnesota, Minneapolis, Minnesota 55455, USA

1 Introduction

Solvent effects can significantly influence the function and reactivity of organic molecules.[1] Because of the complexity and size of the molecular system, it presents a great challenge in theoretical chemistry to accurately calculate the rates for complex reactions in solution. Although continuum solvation models that treat the solvent as a structureless medium with a characteristic dielectric constant have been successfully used for studying solvent effects,[2,3] these methods do not provide detailed information on specific intermolecular interactions. An alternative approach is to use statistical mechanical Monte Carlo and molecular dynamics simulation to model solute–solvent interactions explicitly.[4-8] In this article, we review a combined quantum mechanical and molecular mechanical (QM/MM) method that couples molecular orbital and valence bond theories, called the MOVB method, to determine the free energy reaction profiles, or potentials of mean force (PMF), for chemical reactions in solution. We apply the combined QM-MOVB/MM method to

---

[*] Present address: Department of Chemistry, Western Michigan University, Kalamazoo, MI 49008, USA

ADVANCES IN PHYSICAL ORGANIC CHEMISTRY
VOLUME 38 ISSN 0065-3160   DOI 10.1016/S0065-3160(03)38005-0

three types of nucleophilic substitution reactions and address the question of the dependence of the computed PMF on the choice of the reaction coordinate and its implications on the computed solvent reorganization energy.

The theoretical framework in the present discussion is transition state theory (TST), which yields the expression of the classical rate constant.[9] For a unimolecular reaction, the forward rate constant is given below:

$$k_f = \kappa k_f^{TST} \tag{1}$$

where κ is the transmission coefficient and k_f^{TST} is the TST rate constant. The transmission coefficient accounts for the dynamical correction, which can be calculated by the reactive flux method[10]

$$\kappa(t) = N \langle \dot{X}(0) H[X(t) - X^*] \rangle_{X^*} \tag{2}$$

where X is the reaction coordinate, $\dot{X}(0)$ is the time derivative of X at time $t = 0$ and X^* is the value of the reaction coordinate at the transition state, N is a normalization factor, H is a step function such that it is one when the reactant is in the product side and zero otherwise, and the brackets $\langle \cdots \rangle_{X^*}$ specifies an ensemble average over transition state configurations. Although it is important to consider dynamic effects, the "sobering fact for the theorist" is that the solvent contribution to the free energy of activation often has greater influence on the computed rate constant because of its exponential dependence.[11] Thus, this review focuses on the accurate calculation of solvent effects on the activation free energy.

In classical dynamics, the TST rate constant is the rate of one-way flux through the transition state dividing surface:[9,11]

$$k_f^{TST} = \langle \dot{X} \rangle_{X^*} \, e^{-\beta w(X^*)} / \int_{-\infty}^{X^*} dX \, e^{-\beta w(X)} \tag{3}$$

where $\beta = 1/k_B T$ with k_B being Boltzmann's constant and T the temperature, and $w(X)$ is the PMF along the reaction coordinate X. The frequency for passage through the transition state is given by the average velocity of the reaction coordinate at the transition state, $X = X^*$. Alternatively, equation (3) can be written as

$$k_f^{TST} = \frac{1}{\beta h} e^{-\beta \Delta G^{\neq}} \tag{4}$$

where h is Planck's constant and $\Delta G^{\neq}$ is the molar standard state free energy of activation defined as $\Delta G^{\neq} = w(X^*) - w(X^R)$, with X^R corresponding to the reaction coordinate at the reactant state region.

The PMF is defined as follows:

$$w(X) = -\frac{1}{\beta} \ln \int dX' \, d\mathbf{q} \, \delta[X - X'] e^{-\beta V(X', \mathbf{q})} \tag{5}$$

where $\mathbf{q}$ represents all degrees of freedom of the system except that corresponding to the reaction coordinate, and $V(X', \mathbf{q})$ is the potential energy function. Computationally, the PMF $w(X)$ can be obtained from Monte Carlo and molecular dynamics simulations.[4-8]

It is interesting to note that equation (1) in fact separates the exact classical rate constant into two contributing components, corresponding to the dynamical correction factor and the equilibrium rate constant. Both quantities can be determined from computer simulations, in which the solvent is in thermal equilibrium along the reaction coordinate X. Thus, solvation affects both $\Delta G^{\neq}$ and κ, and these two quantities are not independent of each other, but they are related by the choice of the reaction coordinate X.[11,12] A "better" transition state dividing hypersurface leads to a higher free energy of activation and a smaller number of recrossings, i.e., a larger transmission coefficient. Consequently, in analyzing computational results, it is of interest to examine the effect of using a specific reaction coordinate on the computed PMF.[13-15] In this review, we present a computational approach to investigate the effect of the reaction coordinate on the computed free energies of activation.[14,16]

In the following, we first review the theoretical method and computational techniques. Then, we present the results on three types of nucleophilic substitution reactions in aqueous solution: (i) the Type 1 reaction involving the Cl^- exchange reaction between Cl^- and CH_3Cl, (ii) the Type 3 reaction of Cl^- and $CH_3SH_2^+$, and (iii) the Type 4 charge combination that includes a neutral nucleophile and a cationic substrate, $H_3N + CH_3SH_2^+$. The paper concludes with a summary of the main findings.

2 Methods

There are two main ways of defining the reaction coordinate for condensed phase reactions.[13] A straightforward approach is to use certain geometrical variables that characteristically reflect the chemical process in changing from the reactants to the final products.[17] The solute reaction coordinate, which is denoted by X^R, is akin to studying gas phase reactions through reaction path calculations and it has been successfully applied to numerous organic reactions in solution.[18] When the solute reaction coordinate is used, the solvent environment is in equilibrium with the solute molecule along the entire reaction coordinate.

The second approach is to include explicitly solvent coordinates in the definition of the reaction coordinate because non-equilibrium solvation and solvent dynamics can play an important role in the chemical process in solution.[13] A molecular dynamics simulation study of the proton transfer reaction $[HO\cdots H\cdots OH]^-$ in water indicated that there is considerable difference in the qualitative appearance of the free energy profile and the height of the predicted free energy barrier if the solvent reaction coordinate is explicitly taken into account.[13]

The energy-gap expression, X^S, has often been used as a means to include the solvent degrees of freedom in the definition of the reaction coordinate:[19]

$$X^S = \varepsilon_R - \varepsilon_P \qquad (6)$$

where ε_R and ε_P are, respectively, energies of the reactant and product valence bond (VB) states. Since ε_R and ε_P include solute–solvent interactions, the change in X^S

thus reflects the collective motions of the solvent during the reaction.[13] The reactant and product diabatic states can be approximated using empirical force fields with fixed charges derived from the reactant and product fragments, respectively. This approach has been termed as the empirical valence bond (EVB) method, and it has been applied to a variety of systems.[19] However, the fixed-charge approximation in the definition of diabatic states ignores charge polarization within the molecular fragment due to intra and intermolecular interactions. We summarize recent studies of bimolecular nucleophilic substitution reactions in water by our group, making use of a mixed MOVB approach that provides a quantum mechanical definition of these effective diabatic states.[14,16] Below, we first describe the theoretical background for the definition of effective VB states. Then, we summarize the MOVB method to represent the potential energy surface for chemical reactions. Finally, we present the computational details to obtain free energy profiles.

EFFECTIVE DIABATIC STATES

To express the collective solvent reaction coordinate as in equation (6), it is necessary to define the specific diabatic potential surface for the reactant and product state. This, however, is not a simple task, and there is no unique way of defining such diabatic states. What is needed is a method that allows the preservation of the formal charges of the fragments of reactant and product resonance states. At the same time, solvent effects can be incorporated into electronic structure calculations in molecular dynamics and Monte Carlo simulations. Recently, we developed a block-localized wave function (BLW) method for studying resonance stabilization, hyperconjugation effects, and interaction energy decomposition of organic molecules.[20–23] The BLW method can be formulated to specify the effective VB states.[14]

For a given diabatic resonance state r, we partition the total number of electrons and primitive basis functions into k subgroups, corresponding to a specific form of the Lewis resonance, or VB configuration. For example, the reactant state of the S_N2 reaction between H_3N and $CH_3SH_2^+$ contains two subgroups, one of which (H_3N) has a total charge of zero and the other ($CH_3SH_2^+$) has a total charge of $+1$ e (Fig. 1). Unlike standard Hartree–Fock (HF) theory, molecular orbitals (MOs) in the BLW method are expanded only over atomic orbitals located on atoms within a subgroup.[14,20,21] Therefore, by construction, MOs are no longer delocalized over the entire molecule, but only delocalized within each subgroup of the molecule.

The molecular wave function for resonance state r is

$$\Psi_r = \hat{A}\{\Phi_1^r \Phi_2^r \cdots \Phi_k^r\} \tag{7}$$

where $\hat{A}$ is an antisymmetrizing operator, and Φ_a^r is a successive product of the occupied MOs in subgroup a, $\{\varphi_i^a, i = 1, \ldots, n_a/2\}$

$$\Phi_a^r = \varphi_1^a \alpha \varphi_1^a \beta \cdots \varphi_{n_a/2}^a \beta \tag{8}$$

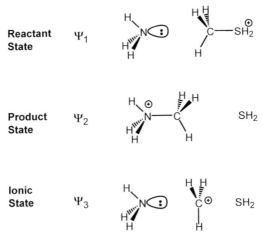

Fig. 1 Schematic representation of three valence bond resonance states for the substitution reaction of ammonia and methylsulfonium ion.

where α and β are electronic spin functions, and n_a is the number of electrons in subgroup a. The MOs in subgroup a are expanded over atomic orbitals located on atoms within that group ($\{\chi_\mu, \ \mu = 1, \dots, m_a\}$)

$$\varphi_j^a = \sum_{\mu=1}^{m_a} c_{j\mu}^a \chi_\mu^a \tag{9}$$

where $c_{j\mu}^a$ are orbital coefficients and m_a is the total number of basis orbitals in the ath subgroup. It is important to note that MOs in the BLW method satisfy the following orthonormal constraints:

$$\langle \varphi_i^a | \varphi_j^b \rangle = \begin{cases} \delta_{ij}, & a = b \\ w_{ij}, & a \neq b \end{cases} \tag{10}$$

where w_{ij} is the overlap integral between two MOs i and j. Equation (10) shows that MOs within each fragment are orthogonal, whereas orbitals in different subgroups are non-orthogonal, a feature of the VB theory.[24,25]

The energy of the localized wave function is determined as the expectation value of the Hamiltonian **H**, which is given as follows:

$$E_r = \langle \Psi_r | \mathbf{H} | \Psi_r \rangle = \sum_{\mu=1}^{} \sum_{\nu=1}^{} d_{\mu\nu} h_{\mu\nu} + \sum_{\mu=1}^{} \sum_{\nu=1}^{} d_{\mu\nu} F_{\mu\nu} \tag{11}$$

where $h_{\mu\nu}$ and $F_{\mu\nu}$ are elements of the usual one-electron and Fock matrix, and $d_{\mu\nu}$ is an element of the density matrix, **D** (Equation 12)[20,26]

$$\mathbf{D} = \mathbf{C}(\mathbf{C}^+ \mathbf{S} \mathbf{C})^{-1} \mathbf{C}^+ \tag{12}$$

where $\mathbf{C}$ is the MO coefficient matrix, $\mathbf{S}$ is the overlap matrix of the basis functions, $\{\chi_\mu^a; a = 1, \ldots, k; \mu = 1, \ldots, m_a\}$. The coefficient matrix for the *occupied* MOs of the BLW wave function has the following form:

$$
\mathbf{C} = \begin{pmatrix} \mathbf{C}^1 & 0 & \cdots & 0 \\ 0 & \mathbf{C}^2 & \cdots & 0 \\ \cdots & \cdots & \cdots & \cdots \\ 0 & 0 & \cdots & \mathbf{C}^k \end{pmatrix} \tag{13}
$$

where the element $\mathbf{C}^a$ is an $n_a/2 \times m_a$ matrix, whose elements are defined in equation (9).

The MOs in equation (9) are optimized using a reorthogonalization technique that has been described by Gianinetti *et al.*,[22] although they can also be obtained using a Jacobi rotation method that sequentially and iteratively optimizes each individual orbital.[20,26]

THE MOLECULAR ORBITAL-VALENCE BOND METHOD

Having defined the effective diabatic states, the wave function for a reacting system, $\Theta[R, X]$, along the entire reaction path can be described by the resonance of these state functions:

$$
\Theta[R, X] = \sum_r a_r \Psi_r[R, X] \tag{14}
$$

where each $\Psi_r[R, X]$ represents a specific diabatic, VB state, and $\Theta[R, X]$ is the adiabatic ground or excited state wave function. To emphasize the fact that the diabatic and adiabatic ground state (as well as excited states) wave functions depend on the geometry of the reactive system $\mathbf{R}$ and the solute–solvent reaction coordinate $\mathbf{X}$, these variables are explicitly indicated in equation (14). The coefficients $\{a_r\}$ in equation (14) are optimized variationally by solving the eigenvalue equation

$$
\mathbf{Ha} = \mathbf{OaE} \tag{15}
$$

where $\mathbf{H}$ is the Hamiltonian matrix, whose elements are defined as $\mathbf{H}_{st} = \langle \Psi_s | \hat{\mathbf{H}} | \Psi_t \rangle$, $\mathbf{a}$ is the state coefficient matrix, and $\mathbf{O}$ is the overlap matrix of non-orthogonal state functions. Evaluation of these matrix elements is straightforward for a given basis set because a number of algorithms have been proposed for solving this problem. Löwdin first described a method on the basis of the Jacobi ratio theorem,[27] whereas Amos and Hall,[28] and King *et al.*[29] developed a bi-orthogonalization procedure for evaluation of matrix elements of non-orthogonal determinant wave functions. In our implementation, we have followed Löwdin's Jacobi ratio strategy.

The method described above has been termed as the MOVB method,[14,16] which represents a combined approach using MO and VB theories. The method perhaps is more conveniently illustrated by a specific example involving the S_N2 reaction of

$H_3N + CH_3SH_2^+ \rightarrow CH_3NH_3^+ + SH_2$. In this case, as in the other two systems described in this paper, the ground state wave function can be adequately described by three effective VB structures, depicted in Fig. 1. These VB states can be represented below:

$$\Psi_1 = \hat{A}\{\Phi(H_3N)\Phi(CH_3SH_2^+)\} \tag{16}$$

$$\Psi_2 = \hat{A}\{\Phi(H_3NCH_3^+)\Phi(SH_2)\}$$

$$\Psi_3 = \hat{A}\{\Phi(H_3N)\Phi(CH_3^+)\Phi(SH_2)\}$$

In equation (16), Ψ_1 and Ψ_2 are the reactant and product state, respectively, and Ψ_3 is an ionic state, which is important for the description of the transition state. More specifically, for the reactant state Ψ_1, $\Phi(H_3N)$ is a product of five doubly occupied MOs localized on the nucleophile ammonia molecule, and $\Phi(CH_3SH_2^+)$ consists of 13 doubly occupied MOs on the fragment of $CH_3SH_2^+$. These spin-orbitals are then antisymmetrized to yield the effective diabatic resonance structure for the reactant state. It is important to note that the partition of the system into different subgroups for each state in equation (16) is used to restrict the region of charge delocalization, irrespective of the specific geometry of the molecular system along the reaction coordinate. Thus, the reactant state will have very high energy in the region corresponding to the product geometry. Conversely, the product state will have high energy in the reactant state region.

A linear combination of the three diabatic states in equation (16) provides a good description of the ground state potential surface in all regions along the reaction coordinate, and the potential energy of the system is obtained by solving the secular equation (15) by diagonalizing the Hamiltonian matrix to yield

$$E_g(X) = \mathbf{a}^+(\mathbf{O}^{-1/2}\mathbf{H}\mathbf{O}^{1/2})\mathbf{a} \tag{17}$$

where, $\mathbf{a}$ is the coefficient matrix, whose elements are defined in equation (14).

FREE ENERGY SIMULATIONS

To include solvent effects in MOVB calculations, we use a combined QM/MM approach, in which the reacting system, or solute, is treated quantum-mechanically and the solvent is represented by a molecular mechanics force field. The effective Hamiltonian in such a combined system is given below:

$$\hat{H} = \hat{H}_{qm}^0 + \hat{H}_{qm/mm} + \hat{H}_{mm} \tag{18}$$

where $\hat{H}_{qm}^0$ is the electronic Hamiltonian for the isolated reactant, $\hat{H}_{qm/mm}$ is the interaction term between the reactant and solvent, and $\hat{H}_{mm}$ is the solvent–solvent interaction energy. Combined QM/MM simulations have been used in numerous applications and have been reviewed previously.[8,30–34] It is interesting to note that the effective Hamiltonian of equation (18) consists of solute–solvent (or QM/MM) interaction terms both in the diagonal, $\mathbf{H}_{ss}$, and the off-diagonal, $\mathbf{H}_{st}$, matrix elements in the MOVB Hamiltonian calculations. Schmitt and Voth found that the effect of the solvent-dependence of the off-diagonal matrix elements is not negligible in studying chemical reactions in solution using an EVB-like approach.[40]

In combined QM-MOVB/MM simulations, the energy-gap reaction coordinate is defined as follows:

$$X^S = E_1(\Psi_1) - E_2(\Psi_2) \tag{19}$$

where $E_1(\Psi_1)$ and $E_2(\Psi_2)$ are the diabatic reactant and product state energies, respectively. In equation (19), the solvent degrees of freedom are included in X^S because the change in solute–solvent interaction energy reflects the collective motions of the solvent molecules as the reaction proceeds.[13,19,36] In this definition, X^S is negative when the system is in the reactant state because the solvent configurations strongly disfavor the product state leading to large positive values in $E_2(\Psi_2)$. X^S changes to a positive value when the system is in the product state because $E_1(\Psi_1)$ will be positive and $E_2(\Psi_2)$ will be negative. Therefore, X^S can be used to monitor the progress of the chemical reaction in solution. Clearly, there is no single reactant structure that defines the transition state, rather, an ensemble of transition states will be obtained from computer simulations.[37–39]

The PMF as a function of X^S is determined by a coupled free energy perturbation and umbrella sampling technique.[5,14,16,41] The computational procedure follows two steps, although they are performed in the same simulation. The first is to use a reference potential E_{RP} to enforce the orientation polarization of the solvent system along the reaction path. A convenient choice of the reference potential, which is called mapping potential in Warshel's work,[13,14,16,42] is a linear combination of the reactant and product diabatic potential energy:

$$E_{RP}(\lambda) = (1 - \lambda)E_1(\Psi_1) + \lambda E_2(\Psi_2) \tag{20}$$

where λ is a coupling parameter that varies from 0, corresponding to the reactant state E_1, to the product state, E_2. Thus, a series of free energy perturbation calculations are executed by moving the variable λ from 0 to 1 to "drive" the reaction from the reactant state to the product state. However, the free energy change obtained using the reference potential, E_{RP}, does not correspond to the adiabatic ground state potential surface. The true ground state PMF is derived from the second step of the computation via an umbrella sampling procedure,[43] which projects the E_{RP} potential on to the adiabatic potential energy surface $E_g(X^S)$:

$$\Delta G(X^S) = \Delta G_{RP}(\lambda) - RT \ln\langle e^{-[E_g(X^S)-E_{RP}(\lambda)]/RT}\rangle_\lambda - RT \ln \rho[X^S(\lambda)] \tag{21}$$

where $\Delta G_{RP}(\lambda)$ is the free energy change obtained in the first step using the reference potential, $E_g(X^S)$ is the adiabatic ground state potential energy at $X^S(\lambda)$, and $\rho[X^S(\lambda)]$ is the normalized distribution of configuration that has a value of X^S during the simulation carried out using $E_{RP}(\lambda)$.

In equation (21), the ground state potential E_g can be either the MOVB adiabatic potential energy or other *ab initio* energies, e.g., the HF, MP2, or DFT energy.

Consequently, the present method is not limited to the MOVB potential energy surface. In our work, we have used both the MOVB and the HF energy as the ground state potential to compare the performance of the method.[14,16]

3 Computational details

MOVB CALCULATIONS

We illustrate the MOVB method by the S_N2 reaction between Cl^- and CH_3Cl, and apply this technique to model substitution reactions. We show that the MOVB method can yield reasonable results for the ground state potential energy surface of the S_N2 reaction both in the gas phase and in solution in comparison with MO and *ab initio* VB calculations. In all calculations, the standard Gaussian 6-31G(d) basis function is used to construct the MOVB wave function.

MONTE CARLO SIMULATION

Statistical mechanical Monte Carlo simulations have been carried out for systems consisting of reactant molecules plus 510–750 water molecules in a periodic cell. Standard procedures are used, including Metropolis sampling and the isothermal–isobaric ensemble (NPT) at 25°C and 1 atm. To facilitate the statistics near the solute molecule, the Owicki–Scheraga preferential sampling technique is adopted with $1/(r^2 + C)$ weighting, where $C = 150$ Å^2. Spherical cutoff distances between 9.5 and 10 Å are used to evaluate intermolecular interactions based on heavy atom separations. For solute moves, all internal geometric parameters including bond lengths, bond angles and dihedral angles are varied, except that the angle between the vectors from the substrate carbon atom to the nucleophile and the leaving group in the first two S_N2 reactions is restricted to be linear. All simulations were maintained with an acceptance rate of *ca.* 45% by using ranges of ± 0.15 Å and 15° for translation and rotation moves of both the solute and solvent molecules. For the internal degrees of freedom, the bond distances are restricted to be ± 0.002 to ± 0.005 Å, bond angles are $\pm 5°$, and the maximum allowed change in dihedral angle is 15°. A series of Monte Carlo free energy simulations are executed, each consisting of $2–3 \times 10^6$ configurations of equilibration and $2–3 \times 10^6$ configurations of averaging. All simulations are performed using a Monte Carlo program developed in our laboratory, which utilizes a locally modified version of the GAMESS program for electronic structure calculations.[44] These simulations were carried out using an IBM-SP system at the Minnesota Supercomputing Institute.

4 Results and discussion

Hughes and Ingold classified nucleophilic substitution reactions into four electrostatic types according to the charge state of the nucleophile and substrate.

This helped to derive the Hughes–Ingold rules for predicting the effect of solvent polarity on the reaction rate.[1] We have used the combined QM-MOVB/MM method to study three nucleophilic substitution reactions in aqueous solution, which are summarized below.

THE TYPE 1 REACTION OF Cl^- + CH_3Cl

A prototype system for computational study of solvent effects is the S_N2 reaction of Cl^- + CH_3Cl, involving an anionic nucleophile and a neutral substrate. The chloride exchange reaction has been extensively studied previously by a variety of theoretical methods.[17,45–50] In this system, as in the other two cases, there are four electrons and three orbitals that directly participate in bond forming and breaking during the chemical reaction.

$$Cl^- + CH_3Cl \rightarrow ClCH_3 + Cl^- \tag{22}$$

The VB wave function for this process, can thus be represented by a linear combination of six Slater determinants corresponding to the VB configurations resulting from this active space. In practice, however, this is not necessary because three determinants, which have very high energies, do not make significant contributions.[5] Consequently, we only need to use three configurations in the VB calculation.[5,14] These VB configurations are listed below:

$$\Psi_1 = \hat{A}\{\Phi(Cl^-)\Phi(CH_3Cl)\} \tag{23}$$

$$\Psi_2 = \hat{A}\{\Phi(ClCH_3)\Phi(Cl^-)\}$$

$$\Psi_3 = \hat{A}\{\Phi(Cl^-)\Phi(CH_3^+)\Phi(Cl^-)\}$$

Here, Ψ_1 and Ψ_2 correspond to the reactant and product state, respectively, and Ψ_3 is a zwitterionic state having two chloride anions separated by a carbocation.

To verify the performance of the MOVB method for studying nucleophilic substitution reactions, we have compared the MOVB data for the chloride exchange reaction with results obtained using other theoretical methods. The gas-phase reaction profile determined using MOVB/6-31G(d) is shown in Fig. 2, along with results obtained from HF/6-31G(d), and *ab initio* valence bond theory (VBSCF). In Fig. 2, the numbers in parentheses specify the number of VB configurations used in the computation, while VBSCF indicates simultaneous optimization of both orbital and configurational coefficients. The term VBCI is used to distinguish calculations that only optimize configuration coefficients (equation 14).

In Fig. 2, the reaction coordinate X^R is the difference between the two C–Cl distances, i.e., $X^R = R_R(C-Cl') - R_P(Cl-C)$, where C–Cl' is the carbon and leaving group distance and Cl–C is the nucleophilic chloride and carbon distance. The double well potential for an S_N2 reaction is clearly characterized by the MOVB method with a binding energy of -9.7 kcal/mol for the ion–dipole complex.[51,52] This may be compared with values of -10.3 kcal/mol from HF/6-31G(d), -10.5 kcal/mol from the G2(+) model,[53] -10.0 kcal/mol from *ab*

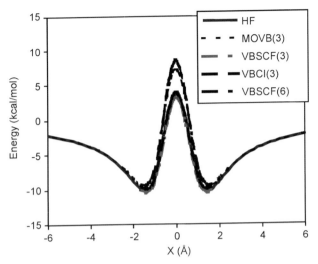

Fig. 2   Computed potential energy profile for the Type 1 S_N2 reaction between chloride ion and methyl chloride in the gas phase.

initio VB, and − 9.4 kcal/mol from a three-configuration VBCI calculation. The experimental binding energy is − 8.6 kcal/mol.[54–56] The barrier height relative to the infinitely separated species is 2.5 kcal/mol from experiment and about 3–4 kcal/ mol from theory. The MOVB and VBCI calculations, which are analogous in that variationally determined VB configurations are used in configuration interaction calculations without further optimizing the orbital coefficients, overestimate the barrier height by about 4–5 kcal/mol in comparison with experiment.[54–56]

The chloride exchange reaction in water has been modeled in Monte Carlo simulations using HF and MOVB methods with the 6-31G(d) basis set.[14] The free energy profile was obtained as a function of the solvent reaction coordinate $X^S = E_1 − E_2$, and the computed PMF for the reaction of $Cl^− + CH_3Cl \rightarrow ClCH_3 + Cl^−$ obtained with the HF/6-31G(d) ground state potential as E_g (equation 21), is shown in Fig. 3.[14] The computed activation free energy in Fig. 3[14] is 26.0 ± 1.0 kcal/mol, which is in excellent agreement with the experimental value (26.6 kcal/mol) and with previous theoretical results. As noted previously, HF/6-31G(d) calculations perform extremely well for the $Cl^− + CH_3Cl$ system, and have been used by Chandrasekhar *et al.* and later by Hwang *et al.* to fit empirical potential functions for condensed phase simulations.[5,45] Thus, the good agreement between MOVB-QM/MM calculations and experiments is not surprising. The striking finding of the large solvent effects, which increase the barrier height by more than 20 kcal/mol is reproduced in the present *ab initio* MOVB calculation.[57] The origin of the solvent effects can be readily attributed to differential stabilization between the ground state, which is charge localized and more stabilized, and the transition state, which is more charge-dispersed and poorly solvated.[5,45] The agreement with experiment for the $Cl^− + CH_3Cl$ S_N2 reaction in water suggest that the MOVB

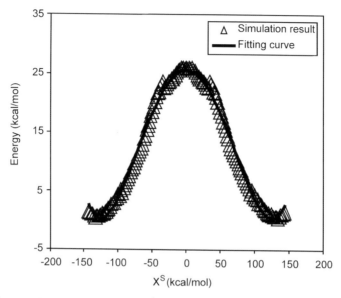

Fig. 3 Computed potential of mean force for the nucleophilic substitution reaction of Cl⁻ and CH_3Cl in water from combined QM-MOVB/MM simulations.

simulation approach along with the use of HF ground state energy can be used to study solvent effects on organic reactions.

THE TYPE 3 REACTION OF $Cl^- + CH_3SH_2^+$

The Type 3 S_N2 reaction between $Cl^- + CH_3SH_2^+$ is interesting because it represents a formal anion–cation recombination through substitution. Because charges are annihilated in forming the transition state, polar solvents will significantly destabilize product formation. Fortunately, the loss in solvation of the two ions is compensated for by electrostatic attractions in bringing the two reactant species into contact. Therefore, the outcome of an S_N2 reaction of Type 3 depends on the balance of Coulomb stabilization and solvent destabilization. The reactant and product diabatic states are defined as follows in MOVB theory:

$$\Psi_1 = \hat{A}\{\Phi(Cl^-)\Phi(CH_3SH_2^+)\} \tag{24}$$

$$\Psi_2 = \hat{A}\{\Phi(ClCH_3)\Phi(SH_2)\}$$

which are used to define the energy-gap reaction coordinate, $X^S = E_1(\Psi_1) - E_2(\Psi_2)$. The PMF has also been computed as a function of the geometrical reaction coordinate, which is defined by $X^R = R_{CS} - R_{ClC}$, where R_{CS} and R_{ClC} are, respectively, distances of S and Cl atoms from the methyl carbon atom.

The free energy profiles along the geometrical reaction coordinate for the reaction of Cl^- and $CH_3SH_2^+$ in water and in the gas phase are shown in Fig. 4. To make it

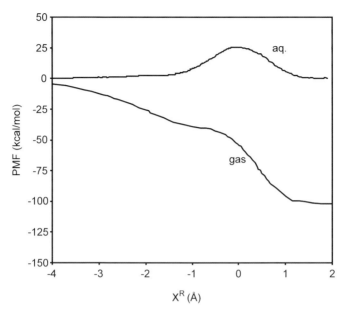

Fig. 4 Computed potential of mean force for the reaction of chloride ion and methylsulfonium ion in water and in the gas phase.

simple for comparison, we have set the zero of energies for both reaction profiles at X^R value of -4 Å, though it should be kept in mind that the Coulomb stabilization is -44.4 kcal/mol at $X^R = -4.5$ Å relative to the infinitely separated ions. Clearly, electrostatic attraction dominates the potential surface in the gas phase as the anionic nucleophile approaches the methylsulfonioum ion. In contrast, a large reaction barrier exists for the aqueous reaction. The free energies of hydration for Cl^- and $CH_3SH_2^+$ ions are -78 and -74 kcal/mol, respectively, which are nearly completely lost at the transition state. The electrostatic attraction energy of about 100 kcal/mol is not sufficient to offset the solvation penalty, leading to a computed free energy barrier of 25.8 kcal/mol. For comparison, a previous theoretical study, employing an *ab initio* MO method coupled with a generalized Born model at the HF/3-21G(d) level, yielded a free energy of activation of 32.4 kcal/mol for the reaction of Cl^- and $CH_3S(CH_3)_2^+$ in water.[58]

The PMF computed using the energy-gap reference potential is shown in Fig. 5, which is obtained from 13 Monte Carlo simulations. Although the general features of the two PMFs for the Type 3 substitution reaction are similar, the predicted free energy barrier along the solvent reaction coordinate is greater than that from the geometrical reaction coordinate by 9 kcal/mol. This suggests that for the charge annihilation process, there is quantitative difference in the computed free energies of activation, perhaps due to a lack of sufficient sampling of solvent configurations as the oppositely charged reactants approach each other in the geometrical mapping procedure. This finding is similar to the study of the proton transfer in the

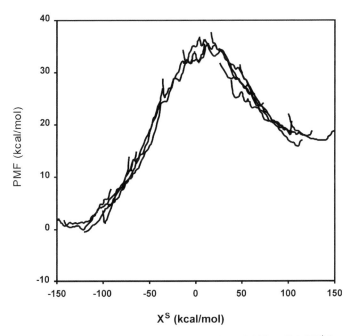

Fig. 5 Computed potential of mean force for the reaction of Cl^- + $CH_3SH_2^+$ in water along the energy-gap solvent reaction coordinate.

[HO$\cdots$H$\cdots$OH]$^-$ system by Muller and Warshel, where significant difference in the predicted activation barrier was noted between geometrical and energy mapping procedures.[13] In both calculations, it seems that it is essential to use a reaction coordinate that explicitly incorporates the solvent coordinates to determine the PMF and free energy barrier.

THE TYPE 4 REACTION OF H_3N + $CH_3SH_2^+$

The substitution reaction between the methylsufonium ion and ammonia involves a neutral nucleophile and a cationic substrate, resulting in dispersal of charges at the transition state. According to the Hughes–Ingold rules, increased solvent polarity will lead to a decrease in the rate of reaction.[1] Thus the free energy of activation is expected to be significantly higher for the reaction in aqueous solution than that in the gas phase. We have investigated the reaction of H_3N + $CH_3SH_2^+$ → $^+H_3NCH_3$ + SH_2 using the combined QM-MOVB/MM simulation method, and the expected solvent effects are confirmed by computer simulations, in accord with experiment.[59]

The three diabatic states used to construct the MOVB wave function have been described in equation (16) earlier to illustrate the computational method. The PMF for the reaction of methylsulfonium ion and ammonia in water have been determined both using the geometrical reaction coordinate (X^R) and the solvent reaction coordinate (X^S) in Monte Carlo simulations. Specifically, the geometrical reaction coordinate is defined as $X^R = R_{CS} - R_{NC}$, where R_{CS} is the distance between the carbon and sulfur atom of the reactant, and R_{NC} is the distance between the nitrogen and carbon atom. The solvent reaction coordinate is described by the energy-gap representation (equation 19). In this study, the 6-31G(d) basis set has been used for the reacting system [$H_3N + CH_3SH_2^+$], and the solvent was represented by the TIP3P model.[60]

Figure 6 compares the PMF for the Type 4 reaction of $H_3N + CH_3SH_2^+$ in water and the energy profile in the gas phase. The gas-phase surface shows a characteristic double-well potential with an ion–dipole complex, having an interaction energy of -9.2 kcal/mol and an overall barrier of 0.4 kcal/mol relative to the separate reactants. The forward reaction is highly exothermic, releasing -34.0 kcal/mol to form methylammonium ion and hydrogen disulfide products. In contrast, the solution-phase reaction has a significant free energy barrier of 16.6 kcal/mol that is induced by solvent effects. Thus, the predicted solvent effect is 16.2 kcal/mol for this reaction. The PMF exhibits a unimodal shape, without the presence of ion–dipole complexes as the nucleophile approaches the substrate to reach the transition state. This feature is reminiscent of the finding for the Type 1 reaction of $Cl^- + CH_3Cl$, both in the gas phase and in water. The increased barrier height in water can be attributed to the change in solvation effects that stabilize more significantly the smaller, charge-localized reactant and product states than the larger, charge-delocalized transition state. An early experimental study of the Type 4 nucleophilic

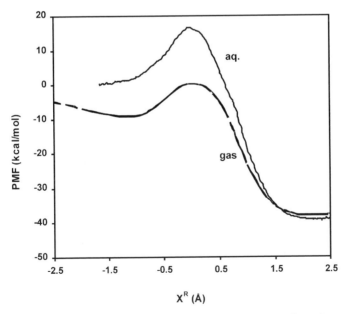

Fig. 6 Computed potential of mean force for the substitution reaction of ammonia and methylsulfonium ion in water and in the gas phase along the solute reaction coordinate.

substitution reaction was reported by Hughes and Whittingham for the reaction of trimethylamine with the trimethylsulfonium ion.[59]

An experimental $\Delta G^{\neq}$ can be derived from the temperature dependence of the second-order rate constant, which yielded a value of 25.9 kcal/mol.[59] Although it appears that this disagrees with the computed free energy of activation (16.6 kcal/mol) for the reaction of $H_3N + CH_3SH_2^+$ in water, the difference actually originates from the intrinsic reactivity of the two reactions. The additional methyl group substitutions both on the nucleophile and substrate raise the gas-phase barrier by 10 kcal/mol to a value of 10.5 kcal/mol at the HF/6-31G(d) level. Taking the methyl substitution effect into account, the computed solvation effect in fact is in accord with experiment,[59] which is about 15 kcal/mol (25.9 − 10.5 kcal/mol).

Figure 7 shows the free energy profile as a function of the energy-gap solvent reaction coordinate, which is compared with the PMF as a function of X^R. The computed $w(X^S)$ also has a unimodal shape and the estimated free energy barrier is 16.1 kcal/mol, in good agreement with the value from Fig. 6. Thus, for the Type 4 reaction, the use of a geometrical and a solvent reaction coordinate does not affect

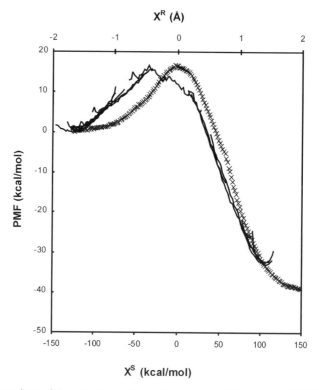

Fig. 7 Comparison of the potentials of mean force for the S_N2 reaction of $H_3N + CH_3SH_2^+$ in water as a function of the solute (upper scale, crossed curve) and solvent (lower scale, solid curves) reaction coordinate, respectively.

the computed free energy of activation significantly, although we note that the Jacobian factor for reaction coordinate transformation has not been corrected.[12] In Fig. 8 is depicted the solute reaction coordinate that has been sampled along the energy-gap, solvent reaction coordinate. At the transition state of the X^S coordinate, the average value of the geometrical reaction coordinate X^R is 0.0 Å, which is identical to that when the PMF is sampled using the geometrical coordinate with the solvent in equilibrium with the solute geometry (Fig. 6). This observation demonstrates that the simulation employing the energy-gap reaction coordinate forces the system to sample the same region of the solute conformational space as that using the geometrical mapping procedure. The near one-to-one correlation between the two reaction coordinates in Fig. 8 suggests that the energy-gap reaction coordinate is an effective approach to monitor the progress of the chemical process, even though geometrical variables of the reactant species are not explicitly used in defining the reaction coordinate.

The use of the energy-gap reaction coordinate allows us to calculate solvent reorganization energies in a way analogous to that in the Marcus theory for electron transfer reactions.[19] The major difference here is that the diabatic states for electron transfer reactions are well-defined, whereas for chemical reactions, the definition of the effective diabatic states is not straightforward. The Marcus theory predicts that

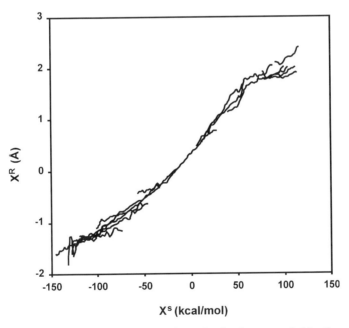

Fig. 8 Variation of the solute reaction coordinate that has been sampled by the energy-gap reference potential for the S_N2 reaction of $H_3N + CH_3SH_2^+$ in water. The figure shows that geometrical variations are closely correlated with the change of the solvent reaction coordinate.

the change of the potential energy for each diabatic state is quadratic in the electron transfer theory, and the solvent reorganization energy can be obtained from the energy of the product diabatic state at the reactant state minimum for a reaction with zero free energy of reaction ($\Delta G^0 = 0$). This approach has been extended to chemical reactions to estimate solvent reorganization energies in water and in enzymes.[5,19,42] However, to obtain parabolic diabatic potential surfaces for chemical reactions, one must make two major assumptions: (1) the solvent response is linear, which is quite reasonable and valid, and (2) the solute charge densities for the reactant and product state (or atomic partial charges) are fixed along the reaction coordinate, which is invalid because of charge polarization due to intramolecular and intermolecular interactions.[14,16] Previous theoretical studies of solvent reorganization energies, employing empirical approaches, obtained parabolic diabatic potential surface. It seems that they are a fortuitous result of the fixed charge assumption. The MOVB method described in this report, however, allows charge polarization within each effective diabatic state, and thus, provides a first-principle examination of the validity of the parabolic potential surfaces for the diabatic states.

Figure 9 illustrates the computed free energies for the reactant and product diabatic states in solution and in the gas phase, and for the solute–solvent interactions. The latter is often used to estimate the solvent reorganization energy

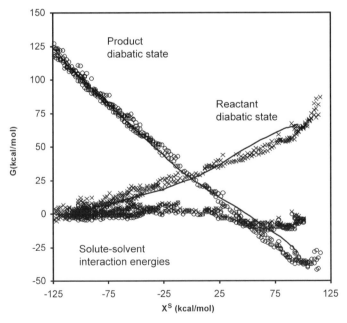

Fig. 9 Computed free energies for the reactant (crosses) and product (circles) diabatic states in solution, and in the gas phase (solid curves). The solute–solvent interaction energies are also shown for the reactant (crosses) and product (circles) diabatic states.

Computational Studies on the Mechanism of Orotidine Monophosphate Decarboxylase

JEEHIUN KATHERINE LEE† and DEAN J. TANTILLO‡

†*Department of Chemistry, Rutgers, The State University of New Jersey, 610 Taylor Road, Piscataway, New Jersey, USA*
‡*Department of Chemistry and Chemical Biology, Cornell University, Ithaca, New York, USA*

This review is dedicated to Ken Houk for his friendship and inspiration.

1 Introduction

Orotidine 5′-monophosphate decarboxylase (ODCase) is a key enzyme in the biosynthesis of nucleic acids, effecting the decarboxylation of orotidine 5′-monophosphate (OMP, **1**) to form uridine 5′-monophosphate (UMP, **2**, Scheme 1).[1,2] The conversion of OMP to UMP is biomechanistically intriguing, because the decarboxylation appears to result, uniquely, in a carbanion (**3**, mechanism i, Scheme 2) that cannot delocalize into a π orbital.[3,4] The uncatalyzed reaction in solution is therefore extremely unfavorable, with a $\Delta G^{\ddagger}$ of

E-mail address: jklee@rutchem.rutgers.edu (J.K. Lee), dt64@cornell.edu (D.J. Tantillo).

ADVANCES IN PHYSICAL ORGANIC CHEMISTRY
VOLUME 38 ISSN 0065-3160 DOI 10.1016/S0065-3160(03)38006-2

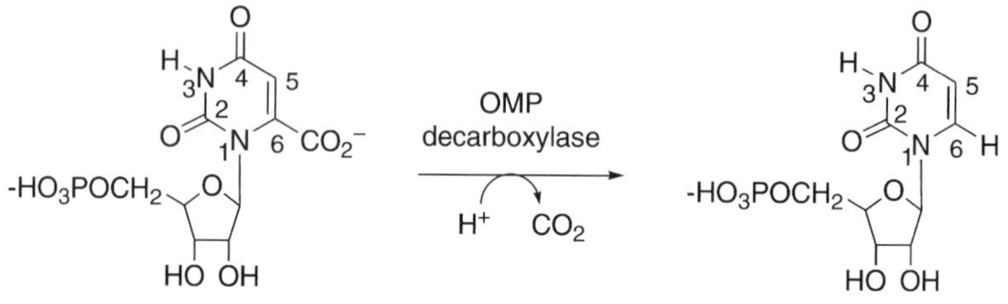

Scheme 1

38.5 kcal mol^{-1} (Fig. 1).[1,5] Remarkably, this activation free energy is lowered to 15.2 kcal mol^{-1} in the ODCase active site (Fig. 1).[1,5] ODCase is thus one of the most proficient enzymes known, with a $k_{cat}/K_m/k_{uncat}$ of 2.0×10^{23} M^{-1}.[1,6] The proficiency is a measure of how effectively the enzyme stabilizes the transition state, and the high proficiency of ODCase indicates that it should be unusually susceptible to transition state analogs as inhibitors.

Because of its essential role in nucleic acid biosynthesis and its unique mechanistic characteristics, ODCase has long been the subject of much study.[6–14] Nonetheless, the catalytic mechanism remains unknown.

Various mechanistic hypotheses have been proposed to explain the fantastic catalysis by ODCase. Particularly prevalent among these hypotheses is proton transfer to the 2-oxygen (the "ylide" mechanism, $1 \rightarrow 4 \rightarrow 5$) or to the 4-oxygen (the "carbene" mechanism, $1 \rightarrow 6 \rightarrow 7$), proposed by Beak and Siegel,[15] and Lee and Houk,[16] respectively (mechanisms ii and iii, Scheme 2). Lee and Houk also proposed that the zwitterionic intermediate formed upon 4-protonation and decarboxylation (7) could be formulated as a heteroatom-stabilized carbene, relatives of which are extremely stable.[17,18] More recently, and very importantly, four different crystal structures of ODCase have been solved and reported by the groups of Ealick and Begley,[19] Short and Wolfenden,[20] Larsen,[21] and Pai and Gao;[22] these structures are of ODCase isolated from *Bacillus subtilis*, *Saccharomyces cerevisiae*, *Escherichia coli* and *Methanobacterium thermoautotrophicum*, respectively. All of these structures are strikingly similar (a representative example is shown in Fig. 2);[23–26] in particular, each active site contains an unusual Asp-Lys-Asp-Lys tetrad. The placement of this tetrad, far from either substrate oxygen but close to the supposed location of the substrate carboxylate, has cast some doubt on the viability of the O-protonation mechanisms and has encouraged new mechanistic proposals involving direct decarboxylation (mechanism i, $1 \rightarrow 3$, Scheme 2), but where catalysis is achieved through ground-state destabilization (repulsion between the substrate carboxylate and a carboxylate of the tetrad that could be relieved as decarboxylation occurs) rather than selective transition state stabilization.[22] Additional mechanistic proposals include proton transfer to the C5 site followed by decarboxylation (mechanism iv, $1 \rightarrow 8 \rightarrow 9$, Scheme 2),[27] and a direct

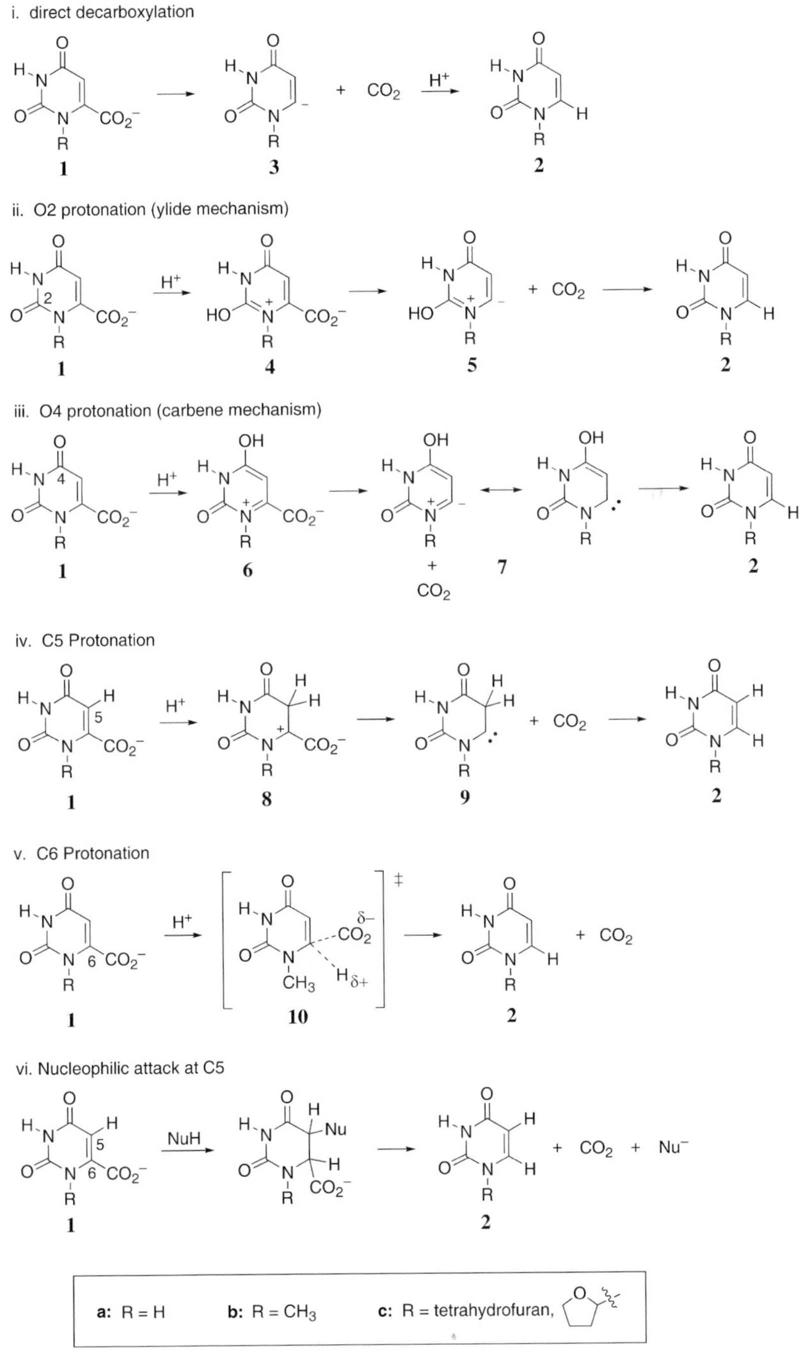

i. direct decarboxylation

ii. O2 protonation (ylide mechanism)

iii. O4 protonation (carbene mechanism)

iv. C5 Protonation

v. C6 Protonation

vi. Nucleophilic attack at C5

a: R = H   b: R = CH₃   c: R = tetrahydrofuran,

Scheme 2

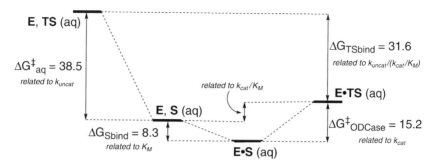

Fig. 1 Experimentally derived binding free energies for the substrate (**S**) and transition state (**TS**) out of aqueous solution to form the ODCase·substrate (**E·S**) and ODCase·transition state (**E·TS**) complexes (ΔG_{Sbind} and ΔG_{TSbind}) and free energies of activation in aqueous solution and ODCase ($\Delta G^{\ddagger}_{aq}$ and $\Delta G^{\ddagger}_{ODCase}$), all in kcal mol^{-1}.

protonation-at-C6/decarboxylation mechanism (mechanism v, **1 → 10 → 2**, Scheme 2).[19] Silverman *et al.* have also suggested a covalent mechanism involving nucleophilic attack at C5 (mechanism vi, Scheme 2), but this was subsequently shown by ^{13}C and D isotope effects to be extremely unlikely.[28,29]

The myriad mechanistic hypotheses have led to a plethora of studies – both experimental and theoretical – aimed at elucidating the ODCase mechanism. This review focuses on those studies which have employed computations as their primary mechanistic tool. These can be divided into two main categories: quantum mechanical studies of small model systems (Section 2), and molecular mechanical studies of the entire enzyme (Section 3). These calculations are often intimately tied to experimental work, and where relevant, experimental studies are described in greater detail.

2   Quantum mechanical studies of OMP decarboxylation

COMPUTATIONAL CHALLENGES

The focus of most computational studies of enzyme catalysis is on locating the structures and relative energies of the stationary points (reactants, intermediates, and transition structures) for the reaction in question in different environments. In principle, this allows one to uncover the origins of enzymatic rate acceleration and/or selectivity by comparing the geometries and relative energies of key points along the reaction coordinate in the gas phase, in a solvent, and in the heterogeneous microenvironment of an enzyme active site.

In general, the structures and relative energies of reactant(s), intermediate(s), and transition structure(s) are first computed for the reaction of interest in the gas phase. This provides a benchmark of the inherent reactivity (activation parameters,

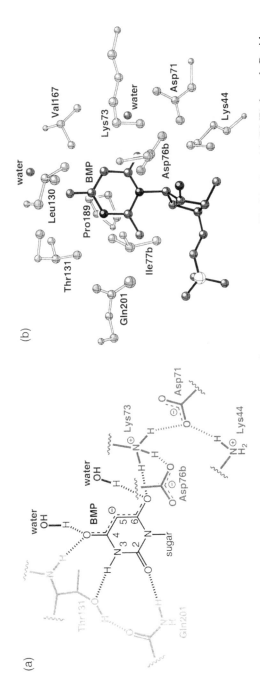

Fig. 2 ODCase active site (pyrimidine-binding subsite) with 1-(5′-phospho-β-D-ribofuranosyl)barbituric acid (BMP) bound. Residues are numbered based on the crystal structure of Larsen and coworkers.[21] (a) Line drawing. Polar uncharged residues are lighter than charged residues. (b) Ball-and-stick drawing based on the crystallographically determined coordinates. Hydrophobic residues are shown in (b) but not in (a).

selectivities) of the substrate(s) in question, to which reactivity in other media can be compared.

Once the gas phase reactivity has been uncovered, the effects of solvation can be explored. There are two families of approaches to this problem, differing in whether or not explicit solvent molecules are used in the calculations. In the first family of approaches – the various continuum or self-consistent reaction field (SCRF) methods[30,31] – the solvent is treated as a homogeneous continuum of a particular dielectric constant, referred to as a reaction field, in which solute molecules (reactants, intermediates, and transition structures) are immersed. In some cases, only single point SCRF calculations are performed, using the structures of stationary points that were computed for the gas phase reaction without further optimization; in other cases, the geometries of these structures are allowed to relax in the presence of the reaction field. Whether or not relaxation is necessary for capturing the effects of solvation depends on the nature of the solvent and solute molecules: in particular, their propensities for forming discrete noncovalent interactions with each other, which can lead to significant changes in geometry and – when the strength of these interactions differs significantly for the reactants and transition structures – activation barriers.

In the second family of approaches, explicit solvent molecules are placed around the gas phase stationary point structures. In some cases, the gas phase geometries are held constant and only the geometries and/or positions of the surrounding solvent molecules are optimized, and in other cases, the structure of the whole system (often called a "supermolecule"[32]) is optimized. The supermolecule approach generally only involves explicit solvent molecules from the first (and occasionally second) solvation shell of the solute.

It is also possible to combine the supermolecule and continuum approaches by using specific solvent molecules to capture the short-range effects (i.e., those involving specific noncovalent interactions between solute and solvent) and a reaction field to treat longer range effects.[33–35] Alternatively, structures along the gas phase reaction coordinate can be immersed in a box of hundreds (or more) of explicit solvent molecules that are treated using force field approaches.[36,37] Each type of method – the SCRF, solvent box, and supermolecule approaches – tests the importance of particular features of the solvent on the reactivity of the solute: dielectric constant, multiple specific classical electrostatic interactions, and specific local directional noncovalent interactions, respectively.

The common approaches for exploring the effects of the enzyme active site on a given reaction are actually quite similar to those used to treat the reaction in solution. In the simplest case, the enzyme environment may be treated using an SCRF method with a dielectric constant corresponding to the interior of the enzyme (various dielectric constants have been used, ranging from 2 to 10).[38] Although this approach involves a gross approximation of the active site environment, it does allow one to assess whether or not specific noncovalent interactions are actually necessary to achieve the observed catalysis.

To quantitate the effects of specific noncovalent interactions, however, the "theozyme" approach[39–42] is often used. This involves calculations in which explicit models of the active site residues surrounding the reactant(s) and transition structure(s) are treated quantum mechanically. In concept, this strategy is quite similar to the supermolecule approach to solvation. In some cases, the positions of the residue models are constrained, usually to coincide with those observed in crystal structures of inhibitor complexes. In other cases, residue structures and positions are fully optimized, although exploring the many conformational minima for such systems can be a daunting task. By comparing the results of both treatments, however, one gets information on whether or not the enzyme has evolved towards preorganization of its active site for ideal reactant or transition state binding. Currently, computations on theozymes involving up to five or so residue models (approximately 15–20 non-hydrogen atoms) can be completed in affordable amounts of time. Theozyme treatments of specific local interactions can also be combined with continuum models of long-range effects. Alternatively, quantum mechanical substrate or theozyme models can be combined with force-field-based treatments of the entire enzyme structure (such approaches are described in Section 3). Again, these combined approaches are similar in concept to those described above for treating solvation effects.

In all of these quantum mechanics-based strategies, calculations of vibrational frequencies can be used to compute zero point energy, thermal, and entropy corrections to the computed internal energies for the various stationary points of interest; the former two corrections allow access to estimates of enthalpies and the latter to free energies. Because of the (statistical mechanical) approximations typically used in entropy calculations of this type, however, computed relative free energies tend to be much less reliable than computed relative enthalpies (which are usually accurate to within several kcal mol^{-1}). It has been suggested based on other types of calculations (see Section 3)[43] and the results of elegant experimental studies[44] that entropies of activation are often quite small and that the origins of enzymatic catalysis are predominantly enthalpic in nature. Yet, because of the relatively small number of systems that have been well-characterized at this point, the importance of entropy for catalysis should be assessed on a case-by-case basis.[45] Computed vibrational frequencies can also be used to predict the effects of isotopic substitution on rates. Since kinetic isotope effects comprise one of the few experimental probes of transition state geometries, they provide an excellent means of validating computational predictions about mechanism (see below for further discussion).

While state-of-the-art quantum mechanical methods are ideally suited for modeling transition structure geometries and activation parameters for gas phase reactions, they are also well-suited for predicting changes to structures and energetics induced by functional groups found in the first solvation shell of the substrate – be they part of solvent molecules or residues in an enzyme active site – and more distant effects of large numbers of solvent molecules or protein residues if they can be captured by an attached method that is computationally affordable (i.e., not requiring quantum mechanical calculations on hundreds or thousands of atoms). Nonetheless, the success of these approaches is somewhat dependent on the particular flavor of quantum mechanics calculations that are used and how the quantum mechanics calculations are combined with continuum or force-field treatments of environmental effects – choices that fall to the chemists carrying out these studies and that should be justified based on appropriate calibration with known experimental observations, such as activation parameters, structures, and isotope effects for the reaction in question and/or closely related systems.

Overall, the quantum mechanical approach to uncovering the sources of enzyme catalysis is one of building up understanding by incrementally adding on models of portions of the enzyme environment (for example, a reaction field or an explicit model of an active site residue) in order to discover how they affect the activation parameters and detailed mechanism of the reaction in question, and whether these components of the enzymatic surroundings produce additive or synergistic effects when combined.

The remainder of Section 2 will discuss the quantum mechanical calculations reported so far on the decarboxylation of orotate derivatives in various environments.

TO PROTONATE OR NOT TO PROTONATE: STUDIES OF THE NAKED SUBSTRATES IN THE GAS PHASE

Three groups have focused on the calculation of the energetics of decarboxylation in the absence of any solvent or active site groups, with only the inclusion of a proton to effect catalysis. Lee and Houk, in 1997, examined the decarboxylation of the simplest parent substrate, orotate (**1a**), in the gas phase.[16] They also studied the effects of protonating the 2- and the 4-oxygen on the energetics. Singleton, Beak, and Lee conducted analogous calculations on 1-methylorotate (**1b**),[46] whereas Phillips and Lee tackled the 1-tetrahydrofuryl-orotate **1c**.[47] Related gas phase studies of proton affinities and acidities of orotate derivatives are also described in this section.

Orotate

Computational approach. Lee and Houk used *ab initio* calculations to ascertain the energetic changes associated with the decarboxylation of orotate in the gas phase.[16] The authors used restricted Hartree–Fock (RHF) calculations with the $6\text{-}31 + G^*$ basis set to conduct full optimizations of the geometries along the decarboxylation reaction coordinate.[48] Energetics were obtained by second-order Møller-Plesset (MP2) energy calculations on the RHF geometries. Density functional methods (Becke3LYP/$6\text{-}31 + G^*$) were also conducted to verify the MP2 values.[49]

Results. The parent decarboxylation reaction of orotate (**1a** → **2a**, Scheme 2) is found to be endothermic in the gas phase: $\Delta H = +43.9$ kcal mol^{-1}. Because there is no barrier to recombination of CO_2 with the carbanion, Lee and Houk equate the endothermicity with the activation enthalpy of the reaction. That is, the reaction is so endothermic that the transition state is late enough to be very close in energy and structure to the product. Therefore, the $\Delta H^{\ddagger}$ of the parent reaction is calculated to be roughly 44 kcal mol^{-1} at MP2/6-31 + G*//RHF-6-31 + G*. The calculated $\Delta G^{\ddagger}$ is 36.4 kcal mol^{-1}. The authors find that the energetics for the uncatalyzed decarboxylation do not appear to be sensitive to solvation; thus, the gas phase values compare favorably to the experimental values in aqueous solution ($\Delta H^{\ddagger} = 44.4$ kcal mol^{-1}; $\Delta G^{\ddagger} = 38.5$ kcal mol^{-1}, Fig. 1).[1] The free-energy barrier for the uncatalyzed reaction has also been calculated using the semi-empirical AM1 method by Gao and coworkers (35.5 kcal mol^{-1}, see Section 3), and is found to be consistent with the Lee–Houk calculations.[22]

Lee and Houk next examined the decarboxylation of the 2-protonated orotate **4a**, which was first proposed by Beak and Siegel in 1976 to be an intermediate for the catalyzed reaction.[15] The $\Delta H^{\ddagger}$ for decarboxylation of **4a** is found to be only 21.6 kcal mol^{-1}, which is 22 kcal mol^{-1} lower than for the uncatalyzed reaction.

Lee and Houk also proposed a modification of the Beak ylide mechanism, suggesting that protonation on the 4-oxygen to yield the stabilized carbene **7** might be a favorable reaction. This "carbene" mechanism (decarboxylation of **6a**, mechanism iii, Scheme 2) is found to have a $\Delta H^{\ddagger}$ of 15.5 kcal mol^{-1}, which is 28 kcal mol^{-1} more favorable than the uncatalyzed reaction. The conclusion from these studies was therefore that both 2- and 4-oxygen protonation would lower the barrier of the decarboxylation, with 4-oxygen protonation (the "carbene" mechanism) being 6 kcal mol^{-1} more favorable than 2-oxygen protonation (the "ylide" mechanism).

1-Methylorotate

Computational approach. Singleton, Beak and Lee explored the pathways of decarboxylation of 1-methylorotic acid **11** via the 2-protonated zwitterion **4b** and the 4-protonated zwitterion **6b** using Becke3LYP calculations with a 6-31 + G* basis set.[46]

11

The free-energy surface was probed by varying the distance "r" between the carboxylate carbon and C6 iteratively, fully optimizing the other coordinates. At each point, the free energy was estimated as $\Delta E - T\Delta S$ by including zero-point energies and entropies based on the unscaled vibrational frequencies. The area around the transition state was found to be very flat, with the energy varying by only about 0.7 kcal mol^{-1} as r was varied from 2.15 to 2.8 Å.

Results. The Singleton–Beak–Lee study focused on comparing experimental and calculated ^{13}C isotope effects on the uncatalyzed decarboxylation of orotic acid (see below). In the course of the study, however, the authors also probed the energetics of decarboxylation of 1-methylorotic acid **11** via the 2-protonated zwitterion **4b** and the 4-protonated zwitterion **6b**. The free-energy barrier for decarboxylation via the 2-protonated zwitterion **4b** at 190 °C is calculated to be 50.5 kcal mol^{-1} starting from **11**. The predicted barrier for decarboxylation via 4-protonation to form the carbene **7b** is significantly lower, at 34.3 kcal mol^{-1}. These values represent the energetic cost in each case for 1-methylorotic acid (**11**) to form a zwitterion, then decarboxylate, and they indicate that the uncatalyzed decarboxylation of 1-methylorotic acid should prefer a 4-protonation pathway by more than 15 kcal mol^{-1}. The authors note that this preference appears to be due, in large part, to the higher basicity of the 4-oxygen versus the 2-oxygen. The authors also conducted solution-phase experiments in heated sulfolane to confirm their computational prediction; these experiments (focusing on isotope effects) are described below.

The barriers just described were calculated with 1-methylorotic acid (**11**) as a reference point to model the uncatalyzed reaction in solution. However, the computed free-energy barriers for decarboxylation of zwitterions **4b** and **6b** are 8.4 and 7.6 kcal mol^{-1}, respectively. This difference of 0.8 kcal mol^{-1} is significantly smaller than the 6 kcal mol^{-1} difference calculated by Lee and Houk for the 2-protonation and 4-protonation pathways. This discrepancy arises from an internal hydrogen bond (**12**) between the N1-H and the carboxylate that artificially stabilizes the O2-protonated zwitterion **4a**, and renders its corresponding decarboxylation barrier too high. When the N1-H is replaced by a methyl, the hydrogen bond is removed, and the ylide and carbene mechanisms become closer in energy; nonetheless, 4-protonation is still favored.

12

1-Tetrahydrofuryl-orotate

Computational approach. Phillips and Lee used *ab initio* calculations to probe the energetics of decarboxylation of the 1-tetrahydrofuryl (THF)-orotate derivative **1c** in the gas phase,[47] utilizing both the GAUSSIAN 94 and GAUSSIAN 98 programs.[48,50] As did Lee and Houk, these authors used RHF calculations with the 6-31 + G* basis set to conduct full optimizations of the geometries along the decarboxylation reaction coordinate. Energetics were obtained by second-order MP2 calculations on the RHF geometries. The starting structures used for these calculations were based on the crystal structure of uridine 5'-monophosphate bound to ODCase in *B. subtilis*,[19,51] and various conformations about the N1–THF bond were explored.

Results. The authors conducted calculations on the 1-THF-orotate (**1c**) system in order to better mimic the actual OMP substrate, where N1 is substituted with a ribose monophosphate (Scheme 1). The enthalpic barrier for the decarboxylation of the 2-protonated zwitterion **4c** is calculated to be 17.0 kcal mol^{-1} whereas the barrier for the 4-protonated zwitterion **6c** is 15.4 kcal mol^{-1}. Consistent with the results obtained by Singleton, Beak and Lee on 1-methylorotate (see above), the carbene pathway is found to be favored slightly over the ylide pathway. Phillips and Lee also obtained calculations on the 1-methylorotate (**1b**) system to assess the effects of switching from a methyl group at N1 to a THF group. The computed barriers for decarboxylation of **4b** and **6b** are 16.9 and 15.1 kcal mol^{-1}, respectively. As in the Singleton–Beak–Lee study, which was conducted at a different theoretical level and focused on free energies rather than enthalpies (see above), these barriers are found to be close in energy, with that for decarboxylation of **6b** being slightly lower. Moreover, these barriers are only slightly different from those computed for **4c** and **6c**, indicating that the switch from methyl to THF did not have a significant effect.

These three main studies of the gas phase behavior of orotate derivatives show that the 4-protonation pathway is always favored over the 2-protonation pathway. When the barriers are calculated relative to a common reference point of orotic acid, as was done in the Singleton–Beak–Lee study, the 4-protonation pathway is favored by a considerable amount, due mostly to the higher basicity of the 4-oxygen over the 2-oxygen in orotate. Still, the 4-protonation pathway also seems to be favored intrinsically, as evidenced by the consistently lower barriers computed for decarboxylation of the 4-protonated zwitterion **6**, regardless of the N1-R group.

Other gas phase studies: proton affinity and acidities

Computed properties of orotate derivatives other than the energetics of decarboxylation have also been published. The computed gas phase proton affinities of the 2- and 4-oxygens of orotate and of C6-deprotonated uracil have been reported by Lee and Houk to be 263 and 274 kcal mol^{-1}, respectively, for orotate O2 and O4, and 285 and 302 kcal mol^{-1} for deprotonated uracil O2 and O4.[16] The authors noted that the greater proton affinity of the 4-oxygen is relevant to the favorability of the 4-protonation pathway. Similar observations were made by Singleton, Beak and Lee and Phillips and Lee.[46,47] Kollman and coworkers recently found that the most basic site of orotate appears to be C5, which is calculated to be 7 kcal mol^{-1} more basic than O4 at MP2/6-31 + G*//HF/6-31 + G*.[27] This translates to a very low energy barrier for decarboxylation of the C5 protonated intermediate: 10 kcal mol^{-1} at MP2/6-31 + G*//HF/6-31 + G*, and 5 kcal mol^{-1} at MP2/cc-pVDZ. SCRF solvation modeling[30,31] using the PCM method increased the barrier to 10, 15, and 21 kcal mol^{-1} using dielectric constants of 2, 4, and 80, respectively. The intriguing proposal that C5-protonation might actually occur in ODCase (mechanism iv, Scheme 2) is discussed further in Section 3.

The proton affinity and acidity of uracil itself (**2a**) has also been the subject of computational investigation, primarily by the groups of Zeegers–Huyskens and Lee.[52–57] Lee and coworkers have also conducted a series of experimental investigations that have established that uracil has four sites that are more acidic than water (N1, N3, C5, and C6) and that O4 is 8 kcal mol^{-1} more basic than O2.[54–56] Gronert and coworkers also conducted clever mass spectrometric experiments that effected decarboxylation of orotate to form the C6-deprotonated uracil (**3a**), which was then used to measure the acidity of the C6-H.[57,58a] Gronert's calculations and experiments, later confirmed by Lee and Kurinovich using different uracil derivatives,[55] established that the C6 site of uracil is quite acidic; with a gas phase acidity (ΔH_{acid}) of ~369 kcal mol^{-1}, this site is as acidic as acetone. Recent experiments in water, however, indicate that the C6-H of 1,3-dimethyl uracil has a pK_a of 34, considerably less acidic than that of acetone (pK_a ~19).[58b]

ADDING ACTIVE SITE COMPONENTS

The logical step following studies of the gas phase behavior of orotate derivatives is to ask what might happen in the actual active site of ODCase. Several groups have attempted to model the active site using theozymes,[39–42] wherein the decarboxylation energetics are calculated quantum mechanically in the presence of one or more relevant functionalities.[16,26,27,59] The conserved Asp-Lys-Asp-Lys tetrad found in all of the crystal structures (rather than the O2–N3–O4 binding region, Fig. 2) has been the focus of most of these theoretical studies because (i) early experiments indicated that an active site lysine residue is essential for catalysis (although not for initial substrate binding),[60] and (ii) the close proximity of an active site aspartate and the substrate carboxylate is a key facet of the ground state destabilization proposal for the origins of catalysis by ODCase.[22]

Methylamine as lysine

Computational approach. Lee and Houk conducted calculations using a methyl-ammonium ion to mimic the key lysine of the enzyme active site.[16] They chose this model because, even though no crystal structures had been solved at the time, a lysine was known to be essential for catalysis.[60] The reaction of orotate + $CH_3NH_3^+$ to form a carbene–methylamine complex was thus examined in various dielectrics using the SCI-PCM SCRF method in GAUSSIAN 94.[30,31,48] Solvation energies computed at the RHF/6-31 + G* level were used to correct gas phase MP2/6-31 + G* energies and obtain ΔH values for reaction in solution.

Results. Lee and Houk were the first to model part of the ODCase active site when they calculated decarboxylation energetics for orotate in the presence of methylammonium ion as a mimic of the key active site lysine. Based on their conclusion that 4-protonation is an energetically favorable pathway (see above), they calculated the energy of reaction of orotate (**1a**) plus $CH_3NH_3^+$ to form a carbene–methylamine complex plus CO_2 (equation 1).

$$\tag{1}$$

This particular reaction model was chosen because the authors proposed that proton transfer should be concerted with decarboxylation. This model reaction is quite exothermic in the gas phase ($-61.9\,\mathrm{kcal\,mol^{-1}}$), but in an environment of low dielectric ($\varepsilon = 4$), as might be expected in an enzyme active site,[38] the $\Delta H^\ddagger$ is a reasonable $17.6\,\mathrm{kcal\,mol^{-1}}$. This barrier is $\sim 25\,\mathrm{kcal\,mol^{-1}}$ less than the $\Delta H^\ddagger$ calculated by these authors for the uncatalyzed decarboxylation of orotate in a water dielectric, which is almost identical to the magnitude of catalysis observed experimentally.[1,6] The authors thus concluded that concerted decarboxylation and proton transfer to the 4-oxygen appears to be a viable catalytic pathway. This particular viewpoint has been challenged by Warshel *et al.*, whose quantum mechanical studies argue against pre-protonation.[61]

Ammonia and methylamine as lysine; formic acid as aspartic acid

Computational approach. Working from the known crystal structures, Kollman and coworkers calculated reaction energy profiles for C5- and C6-protonation pathways using a model system of orotate (**1a**) with an ammonium ion (NH_4^+) to mimic lysine and a neutral formic acid (HCOOH) to mimic aspartic acid, which the authors conclude through molecular dynamics studies should exist in its neutral form (see Section 3).[27] Reaction energy profiles were obtained by constraining the distance between the nearest hydrogen of the ammonium ion and the C5 or C6 of orotate. For each point along the reaction coordinate, all geometric parameters except for the constrained distance were optimized. Optimizations were conducted at HF/6-31 + G*; single point energies were calculated at both HF/6-31 + G* and MP2/6-31 + G*. For some calculations, MP2/cc-pVDZ was also used. Solvation was modeled using the PCM SCRF method.[30,31] Calculations were also conducted in which methylammonium rather than ammonium was used to mimic lysine.

Results. Kollman and coworkers focused on the possibility of C5 or C6 pre-protonation as catalytic pathways to decarboxylation (mechanisms iv and v,

Scheme 2). Initial calculations indicated that C5-protonation is favored over C6-protonation, by about 10 kcal mol^{-1} at MP2/6-31 + G*//HF/6-31 + G* (see above); the authors therefore turned their focus to the C5-protonation mechanism.

To mimic the active site, the energy profile for decarboxylation of **1a** via C5-protonation was computed in the presence of an ammonium ion and formic acid (**13**).

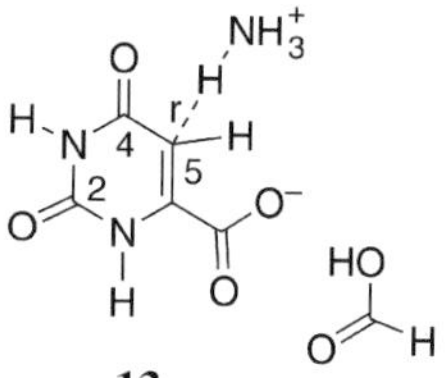

13

The ammonium ion effects the protonation while the formic acid is hydrogen bonded to the orotate carboxylate group. Using this model system, the barrier for C5-protonation was found to be about 25 kcal mol^{-1} at MP2/6-31 + G*//HF/6-31 + G*. Calculations were also conducted using the methylammonium rather than ammonium ion to mimic lysine. In these calculations, formic acid was not present. The energy barrier to C5-protonation with methylammonium is only 18 kcal mol^{-1} at MP2/6-31 + G*//HF/6-31 + G*. Using a larger basis set (MP2/cc-pVDZ) lowers this value to ~ 13 kcal mol^{-1}. Since the experimental $\Delta H^{\ddagger}$ of decarboxylation for ODCase is 11 kcal mol^{-1},[1] C5 pre-protonation appears reasonable on purely energetic grounds.

Solvation was also modeled, using a range of dielectric constant values: 2, 4, and 80. The barriers for C5-protonation by methylammonium become increasingly larger as a higher ε is used; at $\varepsilon = 2$, for example, the energy barrier to protonation increases to 26 kcal mol^{-1}. The authors note, however, that these continuum models, which do not explicitly account for solvent or specific protein residues (see above), may not accurately represent the active site.

Lysine-Aspartate-Lysine-Aspartate tetrad models

Computational approach. Siegbahn and coworkers conducted a series of large calculations using B3LYP in GAUSSIAN 98.[49,50,59] In general, geometries were first optimized using the d95 (double zeta) basis set; then, following this optimization, the energy was calculated using d95 + (2d,2p) (polarization functions added to all atoms and diffuse functions added to the heavy atoms). Test calculations with different basis sets and more demanding theoretical methods (G2MS) were also performed for some simple models to check the accuracy of the B3LYP results.

Several different models of the active site were used, based on the reported crystal structures. In these, the OMP substrate (Scheme 1) was generally modeled as 1-methylorotate (**1b**), but in some calculations, parts of the ribose ring were also included. Lysine residues were modeled as methylamine, aspartate residues were modeled as formate, and glutamine was modeled by acetamide. The part of the protein not explicitly included in these models was treated (using SCRF methods)[30,31] as a homogeneous medium with a dielectric constant of 4.

Results. Siegbahn and coworkers considered three mechanisms: direct (C6)-protonation (mechanism v, $1 \rightarrow 10$, Scheme 2), O2-protonation (mechanism ii, $1 \rightarrow 4 \rightarrow 5$, Scheme 2), and O4-protonation (mechanism iii, $1 \rightarrow 6 \rightarrow 7$, Scheme 2).[59] The direct protonation mechanism was calculated using several different models of the active site wherein some combination of methylamines, aspartates, and/or water was used. The lowest barriers were found for models that involve chains of residues spanning the methylammonium involved in protonating C6 and either O2 or O4 (for example, **14**).

14

The best low-energy models correspond to arrangements that are very unlikely given the crystallographically determined structure of the ODCase active site (Fig. 2). The calculated barriers never drop below 30 kcal mol^{-1}, which is much higher than the experimentally observed barrier for decarboxylation by ODCase ($\Delta G^{\ddagger} = 15$ kcal mol^{-1}, Fig. 1),[1,6] prompting the authors to conclude that direct protonation is not a viable mechanism.

Next, the authors explored the O2 mechanism. In this case, the largest model used (**15**) involved two methylammonium ions and two formate ions to mimic the Lys-Asp-Lys-Asp tetrad, and an acetamide hydrogen-bonded to a water, which interacts with O2, to mimic a glutamine–water bridge found in the Wu–Pai crystal structure.[22]

15

The barrier computed for decarboxylation using this model was 26 kcal mol^{-1}, still somewhat high compared to the experimental value of 15 kcal mol^{-1}.

The O4-protonation mechanism was explored only briefly, due to the fact that the crystal structures do not show any acidic residues in the vicinity of O4 (Fig. 2). The authors did calculate an O4-protonation mechanism using a model in which methylammonium protonates O4 via a bridging water molecule as proposed previously by Houk et al. (16).[23]

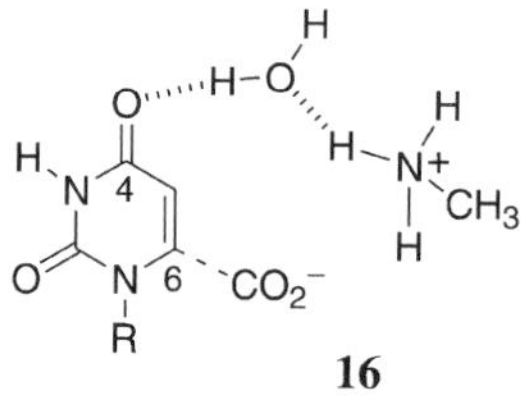

16

This model led to a barrier of ~ 30 kcal mol^{-1}, again much higher than the experimental value.[62] The addition of a formic acid molecule between the water and O4 did not significantly change the computed barrier.

Thus, Siegbahn et al. conclude from their calculations that the most viable mechanism is O2-protonation. Although their calculated barrier of 26 kcal mol^{-1} for this process is still too high, they suggested that larger models (whose sizes are currently prohibitive) may lower this barrier.

ISOTOPE EFFECTS

Isotopic substitution in a molecule can change its rate of reaction significantly and/or shift the balance of equilibria in which it is involved. The magnitude and direction of the isotope effect (IE) – that is, whether a reaction rate or equilibrium concentration increases or decreases upon isotopic substitution – is connected directly to differences in structure between the molecule in question and its transition state for a particular reaction (in the case of a kinetic isotope effect, KIE) or its partners in equilibrium (in the case of an equilibrium isotope effect, EIE). Such effects – whether obtained through experiment or theory or a combination of both – can reveal many intimate details of a mechanistic pathway.

So far, three computational studies of isotope effects related to the ODCase mechanism have been published: Singleton, Beak and Lee used ^{13}C isotope effects to elucidate the mechanism by which the uncatalyzed decarboxylation of orotic acid takes place.[46] Phillips and Lee calculated ^{15}N isotope effects and compared them to known experimental values to show that oxygen-protonation mechanisms are viable for the enzyme-catalyzed process.[47] Kollman and coworkers focused on the ^{15}N isotope effect associated with C5-protonation.[27] Each study is described further below.

Carbon-13 isotope effects and the uncatalyzed decarboxylation of orotic acid

Computational approach. Singleton, Beak, and Lee calculated the ^{13}C isotope effects for each carbon of substrates **4b** and **6b** at several values of *r* along the decarboxylation reaction coordinate, where *r* is the C-CO$_2$ bond distance. The calculations used the program QUIVER[63] with B3LYP frequencies scaled by 0.9614. Tunneling corrections were negligible.[64] Geometries for each structure were optimized at B3LYP/6-31 + G* as described above.

Results. The goal of this study was to elucidate the pathway by which the *uncatalyzed* decarboxylation of 1,3-dimethyl orotic acid in sulfolane proceeds. As described earlier, the authors expected this decarboxylation to proceed via 4-protonation, which is the energetically favored pathway according to their calculations.

The ^{13}C isotope effects for C6 and the carboxylate carbon were found to vary significantly with changing *r*, precluding their use in distinguishing between mechanisms. The isotope effects for C2, C4 and C5, however, varied very little with changing *r*. Of these, only C4 showed substantially different predicted isotope effects for the O2 versus the O4 paths. For the O2-protonation ylide pathway (**11 → 4b → 5b**) a significant secondary isotope effect of 1.006–1.008 is calculated for C4. The isotope effect at C4 for O4-protonation (**11 → 6b → 7b**) is, in contrast, predicted to be unity (1.000). The experimental results for C4, measured by Singleton, are all within error of the predicted values for the decarboxylation *via 4-protonation*. These results provided the first significant experimental support for the theoretically predicted preference for the decarboxylation via O4-protonation.

As noted above, the KIE for the carboxylate carbon was predicted to vary significantly with changing r. The best fit of the experimental and calculated KIEs for the O4-protonated pathway occurs when *r* = 2.65 Å, implying that the transition state for decarboxylation in solution occurs at approximately this C6–CO$_2$ distance. In the gas phase, however, the calculated free energy maximum occurs at *r* = 2.2 Å, and the calculated potential energy maximum is at 2.4 Å. Therefore, the transition state in solution appears later than that in the gas phase. The authors note that one possible explanation for this medium effect could be that the catalytic effectiveness of O4-protonation is lessened in solution because the O4-H group hydrogen bonds to the sulfolane solvent. To test this idea, the pathway for decarboxylation of **6b** was recalculated with the addition of a water molecule hydrogen bonded to the O4-H. The potential-energy maximum did indeed shift later, to *r* = 2.54 Å.

The authors also compared their values to a previously measured ^{13}C isotope effect of 1.043 ± 0.003 for the carboxylate carbon in the *E. coli* ODCase-catalyzed decarboxylation of OMP.[65] This value differs substantially from the experimental value of 1.013 measured by Singleton for the decarboxylation of orotic acid in sulfolane, implying that the uncatalyzed and catalyzed reactions are quite different.

Nitrogen-15 isotope effects and the viability of oxygen protonation as a catalytic path

Computational approach. Phillips and Lee calculated ^{15}N isotope effects[47] using the QUIVER program.[63] The relevant geometries were optimized at B3LYP/ 6-311++G.[49] A scaling factor of 0.96 was used for the frequencies.

Results. The experimental ^{15}N isotope effect at N1 for the decarboxylation of OMP in ODCase (Scheme 1) was measured by Cleland *et al.* to be 1.0068.[66] Comparison of this normal isotope effect with IEs measured for the model compounds picolinic acid (**17**) and *N*-methyl picolinic acid (**18**) led Cleland and coworkers to conclude that the normal IE observed for OMP decarboxylation is indicative of the lack of a bond order change at N1. This conclusion was based on the following reasoning. The IE for the decarboxylation of picolinic acid (**17**) is 0.9955; this inverse value is due to the change in bond order incurred when the proton shifts from the carboxylate group to the N in order to effect decarboxylation (equation 2); the N is ternary in the reactant, but becomes quaternary in the intermediate, which results in the inverse IE. The decarboxylation of *N*-methyl picolinic acid (**18**) involves no such bond order change (equation 3), and the observed normal IE of 1.0053 reflects this.

$$(2)$$

$$(3)$$

Since protonation of the oxygens in OMP may result in some bond order changes at N1 through delocalization, which should manifest themselves in inverse IEs, Cleland and coworkers argued against such mechanisms. Phillips and Lee, however, computed the IEs for N1 in order to figure out whether significant bond order changes actually occur upon oxygen protonation and whether the observed N1 IE for OMP in ODCase really precludes the possibility of O2 and O4-protonation mechanisms.[47]

Phillips and Lee calculated the ^{15}N isotope effect for the decarboxylation of 1-methyl orotate (**1b**) via 2-protonation (**4b**) and via 4-protonation (**6b**). They found that in both cases, the calculated isotope effect is normal: 1.0043 for 2-protonation, and 1.0054 for 4-protonation. An examination of the optimized structures showed clearly that very little bond order change occurs at N1, regardless of which oxygen is protonated. Phillips and Lee also benchmarked their calculations by computing the IEs for protonation of pyridine and for decarboxylation of picolinic acid (**17**) and *N*-methyl picolinic acid (**18**); the results of these calculations are in agreement with the experimental values mentioned above. Therefore, Philips and Lee asserted that the reported ^{15}N-N1 IEs cannot be used to discount the O2 and O4-protonation mechanisms.

As a further step, Phillips and Lee also calculated the ^{15}N decarboxylation isotope effects for the N3 site. For decarboxylation without proton transfer, and for decarboxylation via 2-protonation, the isotope effect is found to be normal (1.0014 and 1.0027, respectively). The 4-protonation pathway, however, has an inverse IE of 0.9949. Therefore, the authors propose that isotope effects at N3 may be useful for distinguishing between these mechanisms.

C5-protonation

Computational methods. Kollman and coworkers calculated the ^{15}N EIEs at N1 for C5-protonation of orotate,[27] using QUIVER.[63] Structures were optimized at HF/6-31 + G* and vibrational frequencies were scaled by 0.8929.

Results. Kollman and coworkers' computed ^{15}N EIEs for C5-protonation of orotate (**1a**) and 1-methylorotate (**1b**) are 0.994 and 0.995, respectively. These inverse IEs indicate that there is some bond order change at N1 upon C5-protonation. The authors point out, however, that this isotope effect must be multiplied by the isotope effect for decarboxylation, which they expect to be normal and large enough to compensate for the inverse IE associated with the C5-protonation step. This would result in a normal IE overall, consistent with the experimental value of 1.0068 measured by Cleland *et al.* (see previous section), and therefore not ruling out their C5-protonation mechanism.

SUMMARY OF QUANTUM MECHANICAL CALCULATIONS AND FUTURE DIRECTIONS

The quantum mechanical studies to date have been steadily building from smaller to larger systems. Initial studies focused on the intrinsic reactivity of orotate (**1a**), 1-methylorotate (**1b**), and 1-tetrahydrofuryl-orotate (**1c**) in the gas phase. These studies examined the effects of proton transfer to the 2- or to the 4-oxygen, and established that pre-protonation on either oxygens significantly lowers the barrier for decarboxylation, with 4-protonation being slightly favored.[16,46,47] Subsequent studies attempted to model ODCase with theozymes composed of key active site functional groups.[16,27,59] Using this approach, Kollman and coworkers added a new dimension to the mechanistic controversy by suggesting that pre-protonation at C5 is also a viable mechanism.[27] The most ambitious quantum mechanical study to date has been undertaken by Siegbahn *et al.*, who have quantum mechanically mimicked not only the Asp-Lys-Asp-Lys active site tetrad, but also active site water molecules and a glutamine found near O2 (see Fig. 2).[59] Based on these large model systems, an O2-protonation mechanism was favored, but still remains far from proven.

In addition to these quantum mechanical studies on the energetics of decarboxylation, several computational studies focused on isotope effects associated with the OMP decarboxylation. A combined experimental and theoretical study of ^{13}C isotope effects in the uncatalyzed decarboxylation of 1,3-dimethyl orotic acid by Singleton, Beak and Lee has provided the first firm evidence for an O4-protonation mechanism.[46] [15]N isotope effects on N1 for the ylide (mechanism ii, Scheme 2) and carbene (mechanism iii, Scheme 2) mechanisms were also computed by Phillips and Lee, who convincingly demonstrated that O2 and O4-protonation are both consistent with the experimentally observed ^{15}N-N1 isotope effect.[47]

Unfortunately, no single mechanism has emerged from these studies as the most likely candidate for the decarboxylation mechanism employed by ODCase. Of the protonation mechanisms, only C6-protonation (mechanism iv, Scheme 2) appears to be consistently discounted,[27,59] and the O2 and O4 pre-protonation mechanisms (mechanisms ii and iii, Scheme 2) still appear to be viable possibilities.[16,46,47,59] The C5-protonation pathway is also a contender.[27]

Nonetheless, there is still hope that quantum mechanical studies may play a key role in deducing the ODCase mechanism. What these studies have shown is that several mechanisms are energetically viable. They have also provided structural models of transition states and their complexes with active site groups that can be used to design experiments for distinguishing between the several mechanisms that remain in the running. One particularly promising experiment that has already been proposed is the measurement of the ^{15}N decarboxylation isotope effects for the N3 site of OMP. Phillips and Lee have made the computational prediction that while decarboxylation via 2-protonation and without pre-protonation should have normal isotope effects (1.0027 and 1.0014, respectively), the 4-protonation pathway should display an inverse IE of 0.9949.[47] Thus, the combination of computationally predicted and experimentally measured IE values may ultimately lead to elucidation of the enzyme mechanism.

3   Free energy computations on OMP decarboxylase

COMPUTATIONAL CHALLENGES

The goal of using free energy calculations to study enzyme catalyzed reactions is to discover *how* a given enzyme increases the rate of a given reaction over its rate in aqueous solution. Two types of strategy are usually applied to this problem.

In one approach, the free energies of binding, out of water into the enzyme active site, of the reactant(s) and transition structure are computed, in order to see if rate acceleration can be explained by selective binding of the transition structure. However, there are several caveats associated with such an approach. First, it must be decided whether to use the same reactant and transition state structures in solution and in the enzyme. If the same structures are used, then the potential for catalysis specifically by selective transition state binding can be quantified. Of course, the actual enzyme-bound structures may be different than those in aqueous solution, and the mechanisms in these two environments may even involve different chemical steps.

The second approach involves directly computing the reaction coordinate for transformation of the enzyme-bound substrate(s) into product(s). Quantum mechanical treatments (see Section 2) are necessary to describe bond-making and breaking processes, however, and such methods are generally too expensive to apply to the whole enzyme–substrate system. Still, if this problem could somehow be circumvented (as has been attempted with QM/MM methods; see below), then assumptions about the structures of species along the reaction coordinate could be avoided.

In the ideal case, both of these approaches are applied to the same reaction. The reaction coordinate in the enzyme is computed directly, and then the binding energies of the reactant(s) and transition structure(s) obtained from these calculations are themselves computed.

In doing these calculations, the first goal is to associate the experimentally determined activation parameters for the enzyme catalyzed reaction with a particular reaction mechanism – ideally, to the exclusion of other alternative mechanisms. In order to accomplish this, the calculations employed must first be able to accurately reproduce the experimental free energy of activation ($\Delta G^{\ddagger}$). In the simplest situation, this will only be possible for one type of mechanism; in practice, however, there may be several mechanistic pathways with similar barriers (i.e., whose difference is smaller than the error bars on the particular type of calculation). When this is the case, computational predictions of other experimentally measurable quantities – such as KIEs (see Section 2) and changes in rate upon mutation of specific protein residues – may allow for differentiation between mechanisms with similar activation parameters.

Even after a particular mechanism is firmly tied to the experimental observations, the job of the theoretician is not complete. Delving deeper into the computational results, by decomposing the $\Delta G^{\ddagger}$ values into chemically meaningful components – the relative importance of enthalpic vs. entropic contributions, electrostatic vs. steric contributions, specific binding interactions vs. solvation effects – is where mechanistic *understanding* arises. Still, the choice of decomposition scheme is somewhat subjective, and therefore the understanding we obtain from such an endeavor is ultimately limited (or perhaps enhanced) by our chemical intuition.

The computational methods

Quantitative prediction of binding energies is still an unresolved computational problem.[67–69] The major difficulty lies in the fact that experimental binding energies arise from differences in the interactions of substrates (reactants, intermediates, transition structures, inhibitors) and enzymes with each other *and with solvent* in the bound and unbound states (as well as solvent–solvent interactions), and the fact that these experimental binding energies include *entropic* as well as enthalpic effects. As of yet, consensus has not been reached on the best methodologies for treating either solvation or entropic effects.[68] Furthermore, when modeling actual chemical reactions in addition to substrate and inhibitor binding, these problems are coupled with the notoriously difficult task of determining the geometry of transition structures for reactions in enzyme active sites.

Current state-of-the-art methods for computing binding free energies of reactants and transition structures and for exploring reaction coordinates for enzyme substrate systems are usually based on MD simulations[69–70] – in the context of free energy perturbation (FEP),[71] empirical valence bond (EVB),[72] and/or linear interaction energy (LIE)[73] schemes. In general, the protein portion of the system is treated with classical molecular mechanics (force-field) methods, and the structures of the substrate portion of the system are derived *separately* from quantum mechanical (QM) calculations (see Section 2). In some cases, however, the structures of the substrate portion of the system are computed directly during the simulation. Such methods allow the structure of the substrate along the reaction coordinate to be determined not only by a QM treatment of its internal preferences for bond lengths, angles, and torsions, but also by the effects of the surrounding protein environment (treated usually with molecular mechanics (MM) methods) – these are the so-called combined QM/MM methods.[74] While the advantages of this sort of calculation are obvious, the drawback of current QM/MM methods is that only computationally inexpensive QM methods (typically semiempirical methods) can be used, which may not give appropriate descriptions of bond-making and breaking. Although these methods cannot yet consistently and quantitatively predict free energies of binding (ΔG_{bind}) and activation free energies ($\Delta G^{\ddagger}$), they can often lead to useful qualitative or semi-quantitative predictions (for example, predicting trends in these quantities for series of related substrates and/or enzymes).

The remainder of Section 3 will discuss the free energy calculations reported so far on ODCase-catalyzed decarboxylation of OMP (Scheme 1).

STEPWISE DECARBOXYLATION AND C6-PROTONATION VIA GROUND STATE DESTABILIZATION?

Computational approach

Gao and coworkers used QM/MM calculations[74] to map out the reaction coordinate and predict the activation free energies for OMP decarboxylation by ODCase and for the decarboxylation of the 1-methylorotate anion (**1b**) in water.[22] Free energies of binding were then computed for structures involved in the decarboxylation using FEP methods.[71]

The aqueous reaction was modeled using Monte Carlo[75] simulations (with umbrella sampling[76]) for a series of points along the $C6–CO_2$ bond-breaking reaction coordinate (the $C6–CO_2$ distances were varied from 1.4 to 7 Å). In each simulation, 1-methylorotate (**1b**) was surrounded by 735 water molecules,[77] and periodic boundary conditions were applied. The electronic structure of 1-methylorotate was treated throughout with the semiempirical AM1 method.[78]

Decarboxylation within the ODCase active site was modeled using MD simulations.[70] The orotate ring system, along with the anomeric carbon of the sugar to which it was attached, were again treated with AM1, while the remainder of the substrate, the enzyme, and a surrounding sphere of water molecules[77] (with radius 24 Å) were treated with the CHARMM22 force field.[79] The generalized hybrid orbital method[80] was used to treat the break between the QM and MM regions at the anomeric carbon of the sugar. This carbon atom was chosen as the QM/MM boundary atom so that the same 1-methylorotate fragment was treated quantum mechanically in both the aqueous and enzyme simulations. Throughout the simulations, two water molecules remained hydrogen-bonded to the orotate substructure (presumably in the vicinity of O2 and O4, although details were not given). Mechanisms involving pre-protonation (mechanisms ii–iv, Scheme 2) or concerted decarboxylation and protonation (mechanism v, Scheme 2) were not allowed in this study, since proton transfer between the QM and MM regions was not possible during the simulations. This effectively limits the mechanistic options to direct decarboxylation followed by proton transfer to obtain UMP (mechanism i, $\mathbf{1} \rightarrow \mathbf{3} \rightarrow \mathbf{2}$, Scheme 2), and only the direct decarboxylation step was simulated in this study.

The free energy changes accompanying the transfer of structures along the reaction coordinate from water to the ODCase active site were then computed using FEP methods.[71] These computations employed a cutoff distance of 14 Å for explicit electrostatic interactions, beyond which a shell (radius 14–16 Å) of dielectric constant 4 was used to approximate the electrostatic properties of the remainder of the protein; the area outside of this shell was treated with a dielectric constant of 78 to represent the electrostatic properties of the surrounding water.

Results

The overall energetics obtained by Gao and coworkers are consistent with previous calculations and experimental values. First, the reasonableness of using AM1, despite its semiempirical nature, is supported by the fact that it predicts an endothermicity of 35.5 kcal mol^{-1} for decarboxylation in the gas phase, which is extremely close to the values predicted previously with more involved computational methods (see Section 2). Second, the predicted activation free energy for decarboxylation in aqueous solution is 37.2 kcal mol^{-1}, while the corresponding experimental value is 38.5 kcal mol^{-1} (Fig. 1).[1] Third, the QM/MM calculations predict a free energy of activation for OMP decarboxylation in ODCase of 14.8 kcal mol^{-1}, while the experimental value is 15.2 kcal mol^{-1} (Fig. 1).[1]

Thus, having confidence in their methodology, Gao and coworkers set about finding a chemically meaningful description of the origins of the observed rate acceleration. They chose to focus on the electrostatic contributions to binding free energies for the reactant and transition structure, and they chose to divide the substrate into a reactive part (the 1-methylorotate substructure) and a "binding block" (the sugar and phosphate groups). The free energy associated specifically with electrostatic interactions between the 1-methylorotate substructure and its surroundings was compared for the reactant and transition state geometries in water and in ODCase. Using this approach, it was determined that the 1-methylorotate group was destabilized upon binding to ODCase by unfavorable electrostatics in both its reactant and transition structure geometries (by 17.8 and 15.6 kcal mol^{-1}, respectively). The small difference between these values (2.2 kcal mol^{-1}) was cited as evidence for a very small contribution to catalysis from selective transition state stabilization. The bulk of the rate acceleration provided by ODCase was ascribed to the large "electrostatic stress" of 17.8 kcal mol^{-1} on the reactive portion of the bound reactant (thought to arise primarily from the proximity of the anionic substrate carboxylate group and that of Asp71; the possibility that this unfavorable interaction is greatly attenuated by the presence of Lys73 and Lys44 was not discussed, however; see Fig. 2). In order for this ground state destabilization mechanism to be valid, the *overall binding* of the 1-methylorotate substructure would have to be unfavorable by this amount, meaning that the interactions of this group with the enzyme are completely accounted for by the computed "electrostatic stress" and that all of the favorable binding energy for OMP as a whole (see Fig. 1) would have to result from an unusually large free energy of binding for the sugar/phosphate binding block. Justification for these assumptions was not provided.

This proposal of a ground state destabilization mechanism for ODCase (this type of mechanism was referred to earlier by Fersht as "electrostatic stress"[81] and by Jencks as the "Circe effect"[82]) sparked considerable controversy.[23–26,83] In some circles it was seen as a prime example of the catalytic power of ground state destabilization,[83] but several groups immediately questioned its validity on the basis of theoretical objections and apparent inconsistencies with biochemical experiments.[23–26]

STEPWISE DECARBOXYLATION AND C6-PROTONATION VIA TRANSITION STATE STABILIZATION?

Computational approach

Warshel and coworkers used EVB[72] and FEP[71] calculations to predict the free energy of activation for ODCase-catalyzed decarboxylation and to examine its origins. This study differs from that of Gao and coworkers in two fundamental ways (besides in the details of the computational methods used in each). First, Warshel and coworkers explored the effects of changing the protonation states of several important residues in ODCase. Second, in some of their simulations, the ammonium group of Lys73 (Fig. 2) was treated quantum mechanically.

After performing *ab initio* and solvation calculations to examine the decarboxylation reaction in water, the free energy surface of the enzyme–catalyzed reaction was explored. An initial ODCase–OMP complex was constructed from the structure of the ODCase–6-azaUMP complex reported by Pai and coworkers,[22] by replacement of 6-azaUMP (**19**) with OMP followed by an MD simulation[70] to allow the system to relax.

19

The free energy surface for the reaction was then calculated using the EVB method[72] in which several valence bond-type configurations (analogous to resonance structures) were used to represent the reacting groups (corresponding to the portion of the system that is treated quantum mechanically; see below) for a given geometric arrangement along the reaction coordinate; the charges of these were allowed to interact with the surrounding environment, and the energies of these "solvated configurations" were then "mixed" to obtain overall free energies. Two models were used for the reacting groups in these calculations: one with only orotate as a reacting group, and another using an orotate/NH_4^+ (to model protonated Lys73) pair. Calculations with these two models differ, therefore, in how the orotate–Lys73 and Lys73–(remainder of ODCase) interactions are treated. Based on the results of simulations of the solution phase decarboxylation of an orotate/NH_4^+ pair, only a stepwise decarboxylation-then-proton transfer mechanism (mechanism i, $1 \rightarrow 3 \rightarrow 2$, Scheme 2) was considered for the enzymatic reaction. Although this treatment is biased against alternative mechanisms, it does allow for a direct evaluation of ODCase's ability to reduce the barrier for such a process. Binding free energies for the reacting groups in their reactant and transition state geometries were then computed using FEP calculations.[71]

In addition, the pK_a values of various residues within the surrounding protein environment were computed using the protein-dipoles Langevin-dipoles model, in a linear response approximation[73] implementation (the PDLD/S-LRA method).[84]

Results

First, different ionization states of the protein residues were examined using PDLD/S-LRA calculations[84] at different protein dielectric constants ranging from 4 to 8.[38] These computations indicated that all four residues in the Asp-Lys-Asp-Lys tetrad are indeed ionized, and that the pK_a of Lys73 was unusually high (16–18), presumably because of its stabilizing interactions with the carboxylate groups that surround it (see Fig. 2). This implies that Lys73 is not the active site residue that was observed experimentally to have a pK_a of approximately 7,[60] but alternative possibilities for the identity of this residue were not suggested.

The decarboxylation reaction coordinate was then explored using the EVB methods described above, assuming, based on the results of solution simulations, that a stepwise decarboxylation-then-proton transfer mechanism (mechanism i, $1 \rightarrow 3 \rightarrow 2$, Scheme 2) is followed. When protonated Lys73 (modeled as NH_4^+) was included as a reactive group (i.e., treated quantum mechanically along with orotate), free energy barriers of $17-24\,kcal\,mol^{-1}$ (depending on the specific ionization state used for the protein) were obtained for decarboxylation. Interestingly, when Lys73 was merely considered as part of the surrounding protein, very similar barriers ($17-23\,kcal\,mol^{-1}$) were obtained. In addition, the activation barrier in aqueous solution was predicted to be $15-22\,kcal\,mol^{-1}$ higher (when using either orotate alone or the orotate/NH_4^+ pair as reacting groups), in reasonable agreement with the experimentally observed value of $23\,kcal\,mol^{-1}$ (see Fig. 1). These calculations suggest that treating Lys73 as a reacting group has no significant effect on catalysis (indicating that either Lys73 is not important for catalysis, or, more probably, that its effects need not be treated quantum mechanically), as long as proton transfer is decoupled from decarboxylation and decarboxylation is rate-determining, as assumed in these calculations.

To evaluate whether the rate acceleration comes from ground state destabilization or transition state stabilization, the electrostatic components of the binding free energies for the reacting groups in their reactant and transition state geometries were evaluated. This was done using FEP methods[71] similar to those used by Gao and coworkers (see above), but which differ in the treatment of long-range electrostatic interactions and in the level of quantum mechanics (ab initio rather than the semiempirical AM1 method) used for computing the charge distributions of the reacting groups. Computed interaction energies for orotate alone were close to zero, while those for the orotate/NH_4^+ reactant pair were very favorable ($20-30\,kcal\,mol^{-1}$). This suggests that the orotate portion of OMP is not significantly destabilized, in contrast to the results of Gao and coworkers (see above). Moreover, the transition state structures are observed to interact more favorably with ODCase (by $10-20\,kcal\,mol^{-1}$, depending on protein ionization state, for orotate alone, and by $17-35\,kcal\,mol^{-1}$ for the orotate/NH_4^+ pair) than the reactant structures, suggesting that ODCase does in fact utilize a transition state stabilization mechanism. This stabilization is ascribed to favorable electrostatic interactions between the substrate and surrounding protein, which increase at the transition state due to an increase in its dipole moment (as negative charge is shifted off the orotate carboxylate and away from Lys73). It is argued that the protonated Lys73 is not hurt by this redistribution of charge because Asp71 and Asp76b (see Fig. 2) are preorganized to interact favorably with it as the carboxylate becomes a CO_2 molecule (in fact, this should also reduce unfavorable Asp–Asp interactions). In solution, it is argued that a significant reorganization energy price must be paid to stabilize the change in dipole moment. The Lys–Asp interactions are taken into account in both types of calculations (i.e., with only orotate treated as a reacting group, and when an orotate/NH_4^+ pair is used), albeit in different ways, explaining why similar results were obtained whether or not Lys73 was treated as a reacting group. This explanation of catalysis has recently been questioned by Siegbahn, however, based on his results using large quantum mechanical models (see above). Nonetheless, the results of Warshel and coworkers indicate that it is possible for a direct decarboxylation mechanism (mechanism i, Scheme 2) to occur in ODCase with a low barrier, and that it is not necessary to invoke a ground state destabilization process to explain the origins of rate acceleration. The viability of pre-protonation or other alternative mechanisms was not directly assessed in this study, however.

Computational approach

Kollman and coworkers used QM-FE[85] (a QM/MM method) and MD/MM-PBSA[86] (a method that involves computing the free energy of a representative selection of "snapshots" from MD trajectories using a molecular mechanics/continuum solvation approach) calculations to explore the possibility of a mechanism involving initial protonation at C5, followed by decarboxylation to produce a carbene and subsequent [1,2]-hydrogen shift (mechanism iv, $1 \rightarrow 8 \rightarrow 9 \rightarrow 2$, Scheme 2). The QM-FE calculations were used to evaluate the effects of the enzyme environment on the gas phase reaction coordinate obtained by *ab initio* calculations (see Section 2), and the MD/MM-PBSA calculations were used to examine the possibility that the ODCase active site is preorganized for C5-protonation. In the QM-FE calculations, only the orotate ring and a methylammonium ion representing Lys73 were treated quantum mechanically. The authors noted that stabilization energies computed by the QM-FE method were unrealistically large and unstable with respect to simulation length and the cutoff distance for computing nonbonded interactions, and also that solute entropic contributions to binding free energies were assumed to be similar for all complexes and were therefore not treated explicitly. These caveats should be kept in mind when drawing conclusions based on the results of this study.

Results

While unable to produce activation parameters that are in quantitative agreement with those measured experimentally, the QM-FE calculations did indicate that the enzyme environment can interact more favorably with a C5-protonated intermediate than with the reactant. Analogous calculations were not reported, however, for alternatively protonated intermediates, so these calculations only suggest the plausibility of C5-protonation, not its predominance over other mechanisms.

The likelihood that Asp71 is ionized at the ODCase active site in the presence of the anionic OMP substrate was explored using a variant of the MM-PBSA method, which involved estimating the pK_a of Asp71 based on the computed difference in deprotonation energy for the neutral form in aqueous solution and in the enzyme–substrate complex (where a dielectric constant of 4 was used). Neutral and anionic structures from the MD simulations were used to compute these deprotonation energies. Using this method, the pK_a of Asp71 in the OMP complex was predicted to be 7.7 ± 2.2 (the intrinsic pK_a for Asp is 4.0), implying that this residue could be protonated in the presence of bound OMP despite the presence of nearby protonated lysine residues (see Fig. 2). Based on this result (and on simulations for complexes of ODCase with 6-azaUMP, **19**), Asp71 was assumed to be in its neutral form for all of the other MD calculations. The neutrality of this Asp was viewed as a means by which the enzyme discourages the protonation of the substrate carboxylate (which would greatly hinder decarboxylation) and thereby reveal a pathway involving protonation of C5. This result also hints that Asp71 may be the catalytic residue responsible for the maximum in V_{max}/K_m observed to occur around pH 7.[60] Proton transfer from Asp71 to the substrate carboxylate group was not examined directly, however. Protonation at an alternative site (C6, O2, or O4) was argued against based on the *ab initio* proton affinities discussed above (see Section 2).

Molecular dynamics simulations on OMP, 2-thio-OMP, and 4-thio-OMP were also undertaken to try and rationalize the experimental observation that thio substitution at the 2-position abolishes catalytic activity while thio substitution at the 4-position only reduces k_{cat} by $\sim 50\%$.[87] It was observed in these simulations that binding of 2-thio-OMP shifts the position of the orotate ring slightly from its computed position in the OMP complex, disrupting the network of hydrogen bonds involving O2 (see Fig. 2). Moreover, C5 and the nitrogen atom of Lys73 are predicted to be in close proximity (≤ 4.6 Å) more often for complexes of OMP and 4-thio-OMP than for the complex of 2-thio-OMP. Similar results were obtained, however, for close approaches of Lys73 and C6. No comment was made on Lys73– O4 distances. In any case, these studies do not involve transition states, and their relevance to the catalytic mechanism is therefore limited.

While this study does apply a variety of computational methods to the problem of ODCase catalysis, the results obtained from the free energy calculations are at their best qualitative. Overall, this report indicates that a C5-protonation mechanism is possible, although its ability to overwhelm alternative mechanisms cannot be asserted.

LOOP DYNAMICS?

Computational approach

Hur and Bruice[88] used MD simulations[70] to examine the structures of ODCase complexes with OMP (**1**) and C6 deprotonated uracil (**3**), and of free OMP in water. Substrate charges used in the simulations were derived from electrostatic potentials computed at the MP2/6-31 + G*//RHF/6-31 + G* level. The two aspartates and two lysines of the active site tetrad were treated as ionized throughout the simulations, and proton transfer to the substrates could not occur during the simulation runs. The initial ODCase–OMP complex was created by modification of the reported structure of ODCase from yeast.[20] The complex with the uracil anion was modeled by eliminating the force that holds the carboxylate to C6. Complexes and free OMP were all immersed in large spheres of water molecules[77] (45 Å radius for the complexes and 25 Å radius for free OMP). Although computed free energies were not reported, the geometries of the substrates and their surroundings were discussed in detail, along with potential implications for proposed catalytic mechanisms.

Results

The optimized geometry of the ODCase–OMP complex was used as evidence against a ground state destabilization mechanism (see above). First, neither Asp group of the tetrad (see Fig. 2) was found to be in close proximity to the carboxylate group of OMP: Asp76b interacts with the OMP carboxylate through a bridging Lys73, and Asp71 interacts with the OMP carboxylate through a bridging water molecule. In addition, structural distortions in OMP upon binding were not observed – the conformation of bound OMP is very similar to one of the major conformers found in the aqueous simulation – suggesting that if destabilization exists it is not manifested in distortions of the substrate geometry. Finally, if strong binding to the nonreactive part of the substrate (the phosphoribosyl substructure) is utilized by the enzyme to force the carboxylate group into a stressed orientation, it is reasonable to expect the region of the active site that binds to the nonreactive part of the substrate to be rather rigid; the simulations show, however, that this portion of the binding site is actually rather flexible.

Arguments against pre-protonation and concerted protonation–decarboxylation mechanisms (mechanisms ii–v, Scheme 2) were also advanced based on these simulations. C5, O2, and O4-protonation were deemed unlikely based on the interactions of these atoms with their surroundings in the ODCase–OMP simulation. C5 was found to be reasonably close to a water molecule (~ 3.7 Å away), which was hydrogen bonded to the OMP carboxylate group, but a similar situation was also observed in the aqueous simulation. O2 was also found to interact with a water molecule, in this case bridged to the phosphate group and a solvent-exposed Arg residue, but it was argued that the series of proton transfers necessary to protonate O2 was unlikely given the relatively small effect on catalysis of mutating this Arg to Ala (a 100-fold drop in k_{cat}, but still a k_{cat}/k_{uncat} of 10^{15}).[89] In addition, it was noted that a short simulation using O2-protonated OMP as substrate indicated that its carboxylate moved away from Lys73 and Asp71, but the stability of this arrangement and its implications for the O2-protonation mechanism are not clear. O4 did form hydrogen bonds to several residues (an NH of a backbone amide group and the sidechains of a Ser and a Cys), but none of these groups is expected to be particularly acidic. No close interactions between O4 and Lys73 were observed. Since actual proton transfers were not allowed during the simulations, pre-protonation (or concomitant protonation and decarboxylation) cannot be ruled out completely.

The optimized structure of the ODCase complex with the uracil anion (**3**) was used to argue for the importance of dynamic effects in transition state stabilization. When the C6–CO_2 distance constraint was removed, CO_2 was released and drifted away from the Asp-Lys-Asp-Lys tetrad, and at the same time, C6 of the resulting uracil anion moved towards Lys73. The Asp-Lys-Asp-Lys tetrad appeared to be quite rigid, however, and so the development of this interaction required movement of the pyrimidine ring system and its attached phosphoribosyl group. This, in turn, caused a major structural change in the surrounding protein: the loop that has been shown to close off the binding site once inhibitors bind[20] shifted its position to further block the phosphoribosyl group from solvent. In the process, this loop also changed its conformation, adopting a β-turn structure. The authors proposed that this conformational change to an ostensibly more stable loop structure may contribute to the stabilization of the ODCase–product and also ODCase–transition state complexes. The magnitude of this dynamic effect remains to be determined, however.

SUMMARY OF FREE ENERGY CALCULATIONS AND FUTURE DIRECTIONS

So far, the several reported studies in which free energy computations have been applied to the mechanism of OMP decarboxylation have not produced an answer to the question of where the rate acceleration provided by ODCase comes from.

At this point it does seem that a direct decarboxylation mechanism (with subsequent non-rate-determining proton transfer) is energetically feasible (mechanism i, Scheme 2). On this point, the studies of Gao and coworkers (see above) and Warshel and coworkers (see above) agree. It also seems clear that the rate acceleration associated with this mechanism does not result from ground state destabilization. This assertion is supported by the calculations of Warshel and coworkers and also by several lines of chemical reasoning.[23–26] One of the main objections to the ground state destabilization hypothesis is that two carboxylates in close proximity (the purported source of the destabilization) would likely not both exist in their anionic forms unless this arrangement is somehow *stabilized*. X-ray structures with inhibitors (for example, Fig. 2) suggest that the substrate and active site carboxylates share a bridging lysine residue (this is also observed in the simulations of ODCase–OMP complexes) which likely serves this role. The simulations of Kollman and coworkers indicate that repulsion between these two carboxylates may also be mitigated by protonation. In other words, although ground state destabilization could, in principle, effectively increase k_{cat}, this would always be at the risk of abolishing the driving force for initial substrate binding (in other words, it may lower k_{cat}/K_m by increasing K_m), and would therefore be unlikely to ever contribute much more than a few kcal mol^{-1} to enzymatic catalysis. Similar opinions have been put forth by Warshel and coworkers,[24–26] who have often emphasized the fact that enzymes evolve to optimize k_{cat}/K_m rather than k_{cat} alone. As shown by Warshel and coworkers (see above), selective transition state stabilization through preorganized electrostatic interactions is a reasonable alternative to ground state destabilization. The suggestion by Bruice and coworkers that large conformational changes during direct decarboxylation may be important for catalysis (see previous subsection) also warrants further study.

Although direct decarboxylation has been shown to be a reasonable mechanism, there have been few free energy calculations on the barriers for other possible mechanisms. In particular, there has yet to be a comprehensive study of the relative energetics of pre-protonation at O2, O4, C5 and concerted protonation–decarboxylation at any of these sites. The suggestion that C5-protonation is possible (see above) is itself intriguing, but more quantitative studies on this mechanism are required to truly establish its viability.

An interesting puzzle has also arisen out of this collection of free energy calculations, which involves the observation, long ago, that ODCase catalysis is pH dependent and maximal around pH 7.[60] It has long been accepted that this observation indicates that there is a key active site residue with a pK_a of ~ 7, and it has often been suggested that this residue is in fact Lys73, its pK_a depressed by a relatively nonpolar microenvironment. Calculations by Warshel and coworkers suggest, however, that the pK_a of this residue is actually *elevated* (to $\sim 16-18$), apparently due mostly to its interactions with Asp71 and Asp76b (see Fig. 2). Analogous calculations by Warshel and coworkers indicate that the pK_a of Asp71 is below 7 (although how much below was not reported), but calculations by Kollman and coworkers (see above) suggest that the Asp71 pK_a may rise as high as 7–9 when the substrate is bound. This opens up the possibility that Asp71 may actually be the residue responsible for the pH dependence of ODCase catalysis. Other possibilities, such as an active site water molecule with an unusually depressed pK_a,[23] are yet to be tested. Further computations and experiments in this area are certainly warranted.

Despite the current lack of consensus on the mechanism employed by ODCase, there is still hope that free energy simulations may eventually converge on an answer. This will be facilitated, we believe, by two things: (1) more comprehensive studies that compare the relative energetics of multiple mechanisms using the same computational methods, and (2) proposals of experiments that will differentiate between mechanisms and computational predictions of their results. For example, it should be possible to use any believable computationally derived model of the ODCase mechanism to predict the effects of specific mutations of enzymatic residues (at least those in the active site) on rate. Stringent tests of this type are necessary not only to reveal problems with computational methods and/or proposed mechanisms, but also to make compelling arguments for the usefulness of theoretical studies and solid chemical reasoning in the world of biology.

4 Overall summary and outlook

Calculations have so far answered some questions related to the decarboxylation of orotic acid derivatives in different media, but these same computational studies have also opened up many additional areas of controversy.

In the gas phase, calculations predict that decarboxylation via O4 pre-protonation should be preferred over decarboxylation via O2 pre-protonation. While this result has not been tested directly in an experimental setting, the greater proton affinity of O4 over O2 in uracil has been established experimentally in the gas phase.

Gas-phase quantum mechanical calculations have also revealed the energetic favorability of C5 pre-protonation, which also awaits experimental verification.

In solution the decarboxylation of 1,3-dimethylorotic acid in sulfolane has been shown through a combination of theoretically predicted and experimentally observed isotope effects to proceed via protonation of the 4-oxygen.

The enzyme mechanism, however, remains elusive. Quantum mechanical models generally disfavor C6-protonation, but O2, O4, and C5-protonation mechanisms remain possibilities. Free energy computations also appear to indicate that C5-protonation is a feasible mechanism, as is direct decarboxylation without pre-protonation; O-protonation mechanisms have yet to be explored with these methods. Controversy remains, however, as to the roles of ground state destabilization, transition state stabilization, and dynamic effects. Because free energy models do take into account the entire enzyme active site, a comprehensive study of the relative energetics of pre-protonation and concerted protonation–decarboxylation at O2, O4, and C5 should be undertaken with such methods. In addition, quantum mechanical isotope effects are also likely to figure prominently in the ultimate identification of the operative ODCase mechanism.

The drive to elucidate this most elusive mechanism thus remains. This is not only a pursuit of interest at an intellectual level, but also biologically and medicinally. As a key enzyme in the biosynthesis of nucleobases, ODCase is a natural target for antitumor agents and genetic diseases, including orotic aciduria;[90] knowledge of the mechanism by which ODCase functions will provide details of the rate-determining transition state structure and should therefore facilitate inhibitor design.

In the long run, it is likely that some combination of quantum mechanical calculations (which tend to build up the enzyme around the reaction) and large-scale free energy calculations (which make predictions regarding the whole enzyme and then endeavor to decompose them into chemically meaningful parts) will play a key role in unraveling the mystery of ODCase.

References

1. Radzicka, A. and Wolfenden, R. (1995). *Science* **267**, 90–93
2. McClard, R.W., Black, M.J., Livingstone, L.R. and Jones, M.E. (1980). *Biochemistry* **19**, 4699–4706
3. Bruice, T.C. and Benkovic, S. (1966). *Bioorganic Mechanisms.* vol. 2. W.A. Benjamin, New York
4. Bender, M.L. (1971). *Mechanisms of Homogeneous Catalysis from Protons to Proteins.* Wiley-Interscience, New York
5. Snider, M.J. and Wolfenden, R. (2000). *J. Am. Chem. Soc.* **122**, 11507–11508
6. For a review of ODCase studies to date, see: Miller, B.G. and Wolfenden, R. (2002). *Ann. Rev. Biochem.* **71**, 847–885. Additional experimental studies not discussed explicitly in this review can be found in references 7–14
7. Creasey, W.A. and Handschumacher, R.E. (1961). *J. Biol. Chem.* **236**, 2058–2063

8. Poulsen, J.-C.N., Harris, P., Jensen, K.F. and Larsen, S. (2001). *Acta Cryst.* **D57**, 1251–1259
9. Bell, J.B., Jones, M.E. and Carter, C.W.J. (1991). *Proteins* **9**, 143–151
10. Wise, E., Yew, W.S., Babbitt, P.C., Gerlt, J.A. and Rayment, I. (2002). *Biochemistry* **41**, 3861–3869
11. Harris, P., Poulsen, J.-C.N., Jensen, K.F. and Larsen, S.J. (2002). *Mol. Bio.* **318**, 1019–1029
12. Song, G.Y., Naguib, F.N.M., el Kouni, M.H. and Chu, C.K. (2001). *Nucleos. Nucleot. Nucl. Acids* **20**, 1915–1925
13. Smiley, J.A., Hay, K.M. and Levison, B.S. (2001). *Bioorg. Chem.* **29**, 96–106
14. Wu, N. and Pai, E.F. (2002). *J. Biol. Chem.* **277**, 28080–28087
15. Beak, P. and Siegel, B. (1976). *J. Am. Chem. Soc.* **98**, 3601–3606
16. Lee, J.K. and Houk, K.N. (1997). *Science* **276**, 942–945
17. Lavorato, D., Terlouw, J.K., Dargel, T.K., Koch, W., McGibbon, G.A. and Schwarz, H. (1996). *J. Am. Chem. Soc.* **118**, 11898–11904, and references therein
18. (a) Arduengo, A.J. (1999). *Acc. Chem. Res.* **32**, 913–921, and references therein; (b) Arduengo, A.J.I., Dias, H.V.R., Dixon, D.A., Harlow, R.L., Klooster, W.T. and Koetzle, T.F. (1994). *J. Am. Chem. Soc.* **116**, 6812–6822, and references therein
19. Appleby, T.C., Kinsland, C., Begley, T.P. and Ealick, S.E. (2000). *Proc. Natl. Acad. Sci. USA* **97**, 2005–2010
20. Miller, B.G., Hassell, A.M., Wolfenden, R., Milburn, M.V. and Short, S.A. (2000). *Proc. Natl. Acad. Sci. USA* **97**, 2011–2016
21. Harris, P., Poulsen, J.-C.N., Jensen, K.F. and Larsen, S. (2000). *Biochemistry* **39**, 4217–4224
22. Wu, N., Mo, Y., Gao, J. and Pai, E.F. (2000). *Proc. Natl. Acad. Sci. USA* **97**, 2017–2022
23. Houk, K.N., Lee, J.K., Tantillo, D.J., Bahmanyar, S. and Hietbrink, B.N. (2001). *ChemBioChem* **2**, 113–118
24. Warshel, A., Florian, J., Strajbl, M. and Villa, J. (2001). *ChemBioChem* **2**, 109–111
25. Warshel, A. and Florian, J. (1998). *Proc. Natl. Acad. Sci. USA* **95**, 5950–5955, and references therein
26. Warshel, A., Strajbl, M., Villa, J. and Florian, J. (2000). *Biochemistry* **39**, 14728–14738
27. Lee, T.-S., Chong, L.T., Chodera, J.D. and Kollman, P.A. (2001). *J. Am. Chem. Soc.* **123**, 12837–12848
28. Silverman, R.B. and Groziak, M.P. (1982). *J. Am. Chem. Soc.* **104**, 6434–6439
29. Smiley, J.A., Paneth, P., O'Leary, M.H., Bell, J.B. and Jones, M.E. (1991). *Biochemistry* **30**, 6216–6223
30. Tomasi, J., Cammi, R. and Mennuci, B. (1999). *Int. J. Quantum Chem.* **75**, 783–803
31. Cramer, C.J. and Truhlar, D.G. (1999). *Chem. Rev.* **99**, 2161–2200
32. Ben-Nun, M. and Levine, R.D. (1995). *Int. Rev. Phys. Chem.* **14**, 215–270
33. Cho, K.H., No, K.T. and Scheraga, H.A. (2000). *J. Phys. Chem. A* **104**, 6505–6509
34. Aleman, C. (1999). *Chem. Phys.* **244**, 151–162
35. Kong, S. and Evanseck, J.D. (2000). *J. Am. Chem. Soc.* **122**, 10418–10427
36. Jensen, C., Liu, J., Houk, K.N. and Jorgensen, W.L. (1997). *J. Am. Chem. Soc.* **119**, 12982–12983
37. For another approach, see: Massova, I. and Kollman, P.A. (2000). *Perspect. Drug Des. Discovery* **18**, 113–135
38. For a recent commentary and leading references, see: Schutz, C.N. and Warshel, A. (2001). *Proteins Struct. Funct. Genet.* **44**, 400–417; See also Simonson, T. and Brooks, C.L., III (1996). *J. Am. Chem. Soc.* **118**, 8452–8458
39. Tantillo, D.J., Chen, J.G. and Houk, K.N. (1998). *Curr. Opin. Chem. Biol.* **2**, 743–750
40. Tantillo, D.J. and Houk, K.N. (2000). *Theozymes and Catalyst Design*, pp. 79–88. Wiley-VCH, Weinhein, Germany

41. Mueller, C., Wang, L.-H. and Zipse, H. (1999). In *ACS Symposium Series 721*. Truhlar, D.G. and Morokuma, K. (eds), pp. 61–73. American Chemical Society, Washington, DC
42. Friesner, R. and Dunietz, B.D. (2001). *Acc. Chem. Res.* **34**, 351–358
43. Villà, J., Strajbl, M., Glennon, T.M., Sham, Y.Y., Chu, Z.T. and Warshel, A. (2000). *Proc. Natl. Acad. Sci. USA* **97**, 11899–11904
44. Wolfenden, R. and Snider, M.J. (2001). *Acc. Chem. Res.* **34**, 938–945
45. For a case where entropy effects are key to catalysis, see: Snider, M.J., Lazarevic, D. and Wolfenden, R. (2002). *Biochemistry* **41**, 3925–3930
46. Singleton, D.A., Merrigan, S.A., Kim, B.J., Beak, P., Phillips, L.M. and Lee, J.K. (2000). *J. Am. Chem. Soc.* **122**, 3296–3300
47. Phillips, L.M. and Lee, J.K. (2001). *J. Am. Chem. Soc.* **123**, 12067–12073
48. Frisch, M.J., Trucks, G.W., Schlegel, H.B., Gill, P.M.W., Johnson, B.G., Robb, M.A., Cheeseman, J.R., Keith, T., Petersson, G.A., Montgomery, J.A., Raghavachari, K., Al-Laham, M.A., Zakrzewski, V.G., Ortiz, J.V., Foresman, J.B., Peng, C.Y., Ayala, P.Y., Chen, W., Wong, M.W., Andres, J.L., Replogle, E.S., Gomperts, R., Martin, R.L., Fox, D.J., Binkley, J.S., Defrees, D.J., Baker, J., Stewart, J.P., Head-Gordon, M., Gonzalez, C. and Pople, J.A. (1995). GAUSSIAN 94, Gaussian, Inc., Pittsburgh, PA
49. (a) Kohn, W., Becke, A.D. and Parr, R.G. (1996). *J. Phys. Chem.* **100**, 12974–12980; (b) Becke, A.D. (1993). *J. Chem. Phys.* **98**, 5648–5652; (c) Becke, A.D. (1993). *J. Chem. Phys.* **98**, 1372–1377; (d) Lee, C., Yang, W. and Parr, R.G. (1988). *Phys. Rev. B* **37**, 785–789; (e) Stephens, P.J., Devlin, F.J., Chabalowski, C.F. and Frisch, M.J. (1994). *J. Phys. Chem.* **98**, 11623–11627
50. Frisch, M.J., Trucks, G.W., Schlegel, H.B., Scuseria, G.E., Robb, M.A., Cheeseman, J.R., Zakrzewski, V.G., Montgomery, J.A., Jr., Stratmann, R.E., Burant, J.C., Dapprich, S., Millam, J.M., Daniels, A.D., Kudin, K.N., Strain, M.C., Farkas, O., Tomasi, J., Barone, V., Cossi, M., Cammi, R., Mennucci, B., Pomelli, C., Adamo, C., Clifford, S., Ochterski, J., Petersson, G.A., Ayala, P.Y., Cui, Q., Morokuma, K., Malick, D.K., Rabuck, A.D., Raghavachari, K., Foresman, J.B., Cioslowski, J., Ortiz, J.V., Baboul, A.G., Stefanov, B.B., Liu, G., Liashenko, A., Piskorz, P., Komaromi, I., Gomperts, R., Martin, R.L., Fox, D.J., Keith, T., Al-Laham, M.A., Peng, C.Y., Nanayakkara, A., Gonzalez, C., Challacombe, M., Gill, P.M.W., Johnson, B., Chen, W., Wong, M.W., Andres, J.L., Gonzalez, C., Head-Gordon, M., Replogle, E.S. and Pople, J.A. (1998). GAUSSIAN 98, Gaussian, Inc., Pittsburgh, PA
51. Begley, T.P., Appleby, T.C. and Ealick, S.E. (2000). *Curr. Opin. Struct. Biol.* **10**, 711–718
52. Chandra, A.K., Nguyen, M.T. and Zeegers-Huyskens, T. (1998). *J. Phys. Chem. A* **102**, 6010–6016
53. Chandra, A.K., Nguyen, M.T., Uchimaru, T. and Zeegers-Huyskens, T. (1999). *J. Phys. Chem. A* **103**, 8853–8860
54. Kurinovich, M.A. and Lee, J.K. (2000). *J. Am. Chem. Soc.* **122**, 6258–6262
55. Kurinovich, M.A. and Lee, J.K. (2002). *J. Am. Soc. Mass Spectrom.* **13**, 985–995
56. Kurinovich, M.A., Phillips, L.M., Sharma, S. and Lee, J.K. (2002). *Chem. Commun.* 2354–2355
57. Gronert, S., Feng, W.Y., Chew, F. and Wu, W. (2000). *Int. J. Mass Spectrom.* **196**, 251–258
58. (a) Feng, W.Y., Austin, T.J., Chew, F., Gronert, S. and Wu, W. (2000). *Biochemistry* **39**, 1778–1783; (b) Sievers, A. and Wolfenden, R. (2002). *J. Am. Chem. Soc.* **124**, 13986–13987
59. Lundberg, M., Blomberg, M.R.A. and Siegbahn, P.E.M. (2002). *J. Mol. Model.* **8**, 119–130
60. (a) Smiley, J.A. and Jones, M.E. (1992). *Biochemistry* **31**, 12162–12168; (b) Smiley, J.A., Paneth, P., O'Leary, M.H., Bell, J.B. and Jones, M.E. (1991). *Biochemistry* **30**, 6216–6223

61. Warshel, A., Strajbl, M. and Florian, J. (2000). *Biochemistry* **39**, 14728–14738
62. Similar results using large model systems have been obtained by Tantillo, D.J. and Houk, K.N., unpublished results
63. Saunders, M., Laidig, K.E. and Wolfsberg, M.J. (1989). *Am. Chem. Soc.* **111**, 8989–8994
64. Bell, R.P. (1980). *The Tunnel Effect in Chemistry.* Chapman and Hall, London
65. Ehrlich, J.I., Hwang, C.-C., Cook, P.F. and Blanchard, J.S. (1999). *J. Am. Chem. Soc.* **121**, 6966–6967
66. Rishavy, M.A. and Cleland, W.W. (2000). *Biochemistry* **39**, 4569–4574
67. (a) For leading references, see: Cozzini, P., Fornabaio, M., Marabotti, A., Abraham, D.J., Kellogg, G.E. and Mozzarelli, A. (2002). *J. Med. Chem.* **45**, 2469–2483; (b) Simonson, T., Archontis, G. and Karplus, M. (2002). *Acc. Chem. Res.* **35**, 430–437
68. (a) Gohlke, H. and Klebe, G. (2002). *Angew. Chem. Int. Ed.* **41**, 2644–2676; (b) van Gunsteren, W.F., Daura, X. and Mark, A.E. (2002). *Helv. Chim. Acta* **85**, 3113–3129
69. Wang, W., Donini, O., Reyes, C.M. and Kollman, P.A. (2001). *Ann. Rev. Biophys. Biomolec. Struc.* **30**, 211–243
70. (a) For leading references, see: Chong, L.T., Duan, Y., Wang, L., Massova, I. and Kollman, P.A. (1999). *Proc. Natl. Acad. Sci. USA* **96**, 14330–14335; (b) Price, D.J. and Brooks, C.L., III (2002). *J. Comp. Chem.* **23**, 1045–1057; (c) Karplus, M. (2002). *Acc. Chem. Res.* **35**, 321–323; (d) Warshel, A. (2002). *Acc. Chem. Res.* **35**, 385–395; (e) Karplus, M. and McCammon, J.A. (2002). *Nat. Struct. Biol.* **9**, 646–652
71. For leading references, see: Kollman, P.A., Massova, I., Reyes, C., Kuhn, B., Huo, S., Chong, L., Lee, M., Lee, T., Duan, Y., Wang, W., Donini, O., Cieplak, P., Srinivasan, J., Case, D.A. and Cheatham, T.E., III (2000). *Acc. Chem. Res.* **33**, 889–897
72. For leading references, see: Åqvist, J. and Warshel, A. (1993). *Chem. Rev.* **93**, 2523–2544
73. (a) Åqvist, J., Medine, C. and Samuelsson, J.E. (1994). *Protein Engng* **7**, 385–391; (b) Åqvist, J., Luzhkov, V.B. and Brandsdal, B.O. (2002). *Acc. Chem. Res.* **35**, 358–365
74. (a) Lee, T.-S., Massova, I., Kuhn, B. and Kollman, P.A. (2000). *J. Chem. Soc., Perkin Trans.* **2**, 409–415; (b) Monard, G. and Merz, K.M., Jr (1999). *Acc. Chem. Res.* **32**, 904–911; (c) Åqvist, J. and Warshel, A. (1993). *Chem. Rev.* **93**, 2523–2544; (d) Warshel, A. (1991). *Computer Modeling of Chemical Reactions in Enzymes and Solutions.* Wiley, New York; (e) Related approaches include the use of force field methods designed to locate transition structures directly (Esterowicz, J.E. and Houk, K.N. (1993). *Chem. Rev.* **93**, 2439–2461 and Jensen, F. and Norrby, P.-O. (2003). *Theor. Chim. Acc.* **109**, 1–7) and the transition-state docking approach: Tantillo, D.J. and Houk, K.N. (2002). *J. Comp. Chem.* **23**, 84–95
75. See, for example: Leach, A.R. (1996). *Molecular Modelling: Principles and Applications*, Chapter 7. Addison Wesley Longman, Edinburgh Gate
76. See, for example: Leach, A.R. (1996). *Molecular Modelling: Principles and Applications*, pp. 498–499. Addison Wesley Longman, Edinburgh Gate
77. TIP3P water molecules were used. See: Jorgensen, W.L., Chandrasekhar, J., Madura, J.D., Impey, R.W. and Klein, M.L. (1983). *J. Chem. Phys.* **79**, 926–935
78. Dewar, M.J.S., Zoebisch, E.G., Healy, E.F. and Stewart, J.J.P. (1985). *J. Am. Chem. Soc.* **107**, 3902–3909
79. MacKerell, A.D., Jr, Bushford, D., Bellott, M., Dunbrack, R.L., Evanseck, J.O., Field, M.J., Fischer, S., Gao, J., Guo, H., Ha, S., Joseph-McCarthy, D., Kuchnir, L., Kuczera, K., Lau, F.T.K., Mattos, C., Michnick, S., Ngo, T., Nguyen, D.T., Prodhom, B., Reiher, W.E., III, Roux, B., Schlenkrich, M., Smith, J.C., Stote, R., Straub, J., Watanabe, M., Wiorkiewicz-Kuczera, J., Yin, D. and Karplus, M. (1998). *J. Phys. Chem. B* **102**, 3586–3616
80. Gao, J., Amara, P., Alhambra, C. and Field, M.J. (1998). *J. Phys. Chem. A* **102**, 4714–4721

81. Fersht, A. (1999). *Structure and Mechanism in Protein Science*, pp. 372–375. Freeman, New York
82. Jencks, W.P. (1975). *Adv. Enzymol. Relat. Areas Mol. Biol.* **43**, 219–410
83. (a) Editorial by Chin, G. (2000). *Science* **288**, 401; (b) Rouhi, A.M. (2000). *Chem. Engng News*, **75**, 42–52
84. Lee, F.S., Chu, Z.T. and Warshel, A. (1993). *J. Comput. Chem.* **14**, 161
85. (a) Stanton, R., Perakyla, M., Bakowies, D. and Kollman, P. (1998). *J. Am. Chem. Soc.* **120**, 3448–3457; (b) Kuhn, B. and Kollman, P. (2000). *J. Am. Chem. Soc.* **122**, 2586–2596; (c) Lee, T.-S., Massova, I., Kuhn, B. and Kollman, P.A. (2000). *J. Chem. Soc., Perkin Trans.* **2**, 409–415; (d) Kollman, P.A., Kuhn, B., Donini, O., Perakyla, M., Stanton, R. and Bakowies, D. (2000). *Acc. Chem. Res.* **34**, 72–79
86. Srinivasan, J., Cheatham, T., Cieplak, P., Kollman, P. and Case, D. (1998). *J. Am. Chem. Soc.* **120**, 9401–9409
87. Shostak, K. and Jones, M. (1992). *Biochemistry* **31**, 12155–12161
88. Hur, S. and Bruice, T.C. (2002). *Proc. Natl. Acad. Sci.* **99**, 9668–9673
89. Miller, B.G., Snider, M.J., Short, S.A. and Wolfenden, R. (2000). *Biochemistry* **39**, 8113–8118
90. Suttle, D.P., Becroft, D.M.O. and Webster, D.R. (1989). In *The Metabolic Basis of Inherited Diseases*, Scriver, C.R. (ed.), pp. 1095–1126. McGraw Hill, New York

Cumulative Index of Authors

Cumulative Index of Titles

221

Subject Index